21世纪高等院校移动开发人才培养规划教材
21Shiji Gaodeng Yuanxiao Yidong Kaifa Rencai Peiyang Guihua Jiaocai

Android 应用程序设计教程

李华忠 梁永生 刘涛 主编

Android
Application Design

人民邮电出版社
北京

图书在版编目（CIP）数据

Android应用程序设计教程 / 李华忠，梁永生，刘涛主编. -- 北京：人民邮电出版社，2013.11（2017.7重印）
21世纪高等院校移动开发人才培养规划教材
ISBN 978-7-115-32968-4

Ⅰ. ①A… Ⅱ. ①李… ②梁… ③刘… Ⅲ. ①移动终端－应用程序－程序设计－高等学校－教材 Ⅳ.
①TN929.53

中国版本图书馆CIP数据核字(2013)第204434号

内 容 提 要

全书共分成 8 章，主要内容包括 Android 开发环境构建、Android 屏幕布局、Android 控件 Widgets、Android 的图形用户界面、Android 数据存储、Android 多媒体以及两个综合应用了本课程核心知识和关键技术的综合实训项目（手机乐游和基于 Android 的手机定位）。

本书符合教学规律和课堂要求，很好地反映了嵌入式和移动互联等行业出现的 Android 方面的新知识、新技术、新方法和新应用，能解决高校 Android 课程教学面临的迫切问题，既可作为高等院校 Android 应用程序设计课程的教材，也可作为移动开发爱好者的自学参考书。

◆ 主　编　李华忠　梁永生　刘　涛
　　责任编辑　王　威
　　责任印制　杨林杰

◆ 人民邮电出版社出版发行　　北京市丰台区成寿寺路 11 号
　　邮编　100164　电子邮件　315@ptpress.com.cn
　　网址　http://www.ptpress.com.cn
　　北京京华虎彩印刷有限公司印刷

◆ 开本：787×1092　1/16
　　印张：15　　　　　　　　2013 年 11 月第 1 版
　　字数：383 千字　　　　　2017 年 7 月北京第 5 次印刷

定价：39.80 元（附光盘）

读者服务热线：(010)81055296　印装质量热线：(010)81055316
反盗版热线：(010)81055315
广告经营许可证：京东工商广登字 20170147 号

前　言

Android 是一个由谷歌和开放手机联盟（Open Handset Alliance）开发与领导，以 Linux 为基础的，完整、开放、免费的手机平台。它由应用程序、应用程序框架、系统库、Android 运行时以及 Linux 内核这 5 部分组成。基于 Android 的应用程序设计以面向对象 Java 语言实现的应用程序框架为基础，易学、易用，从而极大地降低了在手机和平板电脑等终端设备上开发移动互联应用程序的难度，使 Apps 应用程序开发的效率大大提高，已经成为世界上主流移动应用程序开发平台之一。

目前，我国很多院校的计算机软件、移动互联、嵌入式和物联网等相关专业，都将"Android 应用程序开发"作为一门重要的专业支撑课程。为了帮助院校老师能够比较全面、系统地讲授这门课程，使学生能够熟练地使用 Android 来进行移动互联软件开发，我们几位长期在院校从事 Android 教学的教师和企业工程师，共同编写了这本《Android 应用程序设计》。

我们对本书的体系结构做了精心的设计，按照 Android 平台的技术体系结构和项目内容，如 Android 开发环境、屏幕布局、控件、图形用户界面、Android 数据存储和 Android 多媒体等项目，设计多个学习情境。每个学习情境又结合知识体系和实践技能细化为若干个子学习情境，由浅入深，实用性强。最后，结合移动互联应用实际情况，安排两个综合实训项目，在提高学生应用技能的同时，强化项目驱动，实施"工学结合"，提高理论教学和实践教学质量，充分满足了对高职院校学生教学和自学需求。在内容编写方面，我们注意难点分散、循序渐进；在文字叙述方面，我们注意言简意赅、重点突出；在实例选取方面，我们注意实用性强、针对性强。

本书每章都附有一定数量的习题，可以帮助学生进一步巩固基础知识；本书每章还附有实践性较强的项目实施，可以供学生上机操作时使用。本书配备了 PPT 课件、源代码、习题答案、教学大纲、课程设计等丰富的教学资源，任课教师可到人民邮电出版社教学服务与资源网（www.ptpedu.com.cn）免费下载使用。本书的参考学时为 64 学时，其中实践环节为 20 学时，各章的参考学时参见下面的学时分配表。

章节	课程内容	学时分配	
		讲授	实训
第1章	Android开发环境构建	4	2
第2章	Android屏幕布局	4	2

续表

章节	课程内容	学时分配	
		讲授	实训
第3章	Android控件Widgets	6	2
第4章	Android的图形用户界面	6	2
第5章	Android数据存储	6	2
第6章	Android多媒体	6	2
综合实训1	手机乐游项目	6	4
综合实训2	基于Android的手机定位项目	6	4
课时总计		44	20

  本书由深圳信息职业技术学院李华忠撰写第1章至第5章，梁永生撰写第6章和综合实训1，深圳市优科赛服网络科技有限公司资深工程师刘涛撰写综合实训2，并负责本书参考代码的测试工作。

  由于移动开发技术发展日新月异，加之我们水平有限，书中难免存在错误和不妥之处，敬请广大读者批评指正。

编者

2013 年 9 月

目　录

第 1 章　Android 开发环境构建 ………… 1
1.1　项目导引 ……………………………… 1
1.2　项目分析 ……………………………… 1
1.3　技术准备 ……………………………… 2
1.3.1　Android 系统架构 ……………… 2
1.3.2　开发环境搭建 …………………… 5
1.3.3　创建 Android 应用程序 ……… 14
1.3.4　解析 Android 应用程序框架 … 19
1.4　项目实施 …………………………… 24
1.5　技术拓展：调试程序方法（DDMS 和 Logcat） ……………………………… 25
1.6　本章小结 …………………………… 29
1.7　强化练习 …………………………… 29

第 2 章　Android 屏幕布局 ……………… 30
2.1　项目导引 …………………………… 30
2.2　项目分析 …………………………… 30
2.3　技术准备 …………………………… 31
2.3.1　线性布局（LinearLayout） …… 31
2.3.2　相对布局（Relative Layout） … 34
2.3.3　表单布局（Table Layout） …… 37
2.3.4　单帧布局（Frame Layout） …… 42
2.3.5　坐标布局（AbsoluteLayout） … 43
2.3.6　切换卡（Tab Widget） ………… 45
2.4　项目实施 …………………………… 49
2.5　技术拓展：<include>和自定义控件 … 54
2.6　本章小结 …………………………… 61
2.7　强化练习 …………………………… 61

第 3 章　Android 控件 Widgets ………… 62
3.1　项目导引 …………………………… 62

3.2　项目分析 …………………………… 62
3.3　技术准备 …………………………… 65
3.3.1　知识点 1：文本框（TextView） … 65
3.3.2　知识点 2：编辑框（EditText） … 69
3.3.3　知识点 3：按钮（Button）和图片按钮（ImageButton） …… 72
3.3.4　知识点 4：复选框（Check Box）和单选按钮（Radio Button） ……………………………… 74
3.3.5　知识点 5：数字时钟与模拟时钟（AnalogClock，DigitalClock） · 83
3.3.6　知识点 6：日期与时间（DatePicker，TimePicker） …… 86
3.4　项目实施 …………………………… 91
3.5　技术拓展 …………………………… 95
3.6　本章小结 …………………………… 100
3.7　强化练习 …………………………… 100

第 4 章　Android 的图形用户界面 … 102
4.1　项目导引 …………………………… 102
4.2　项目分析 …………………………… 102
4.3　技术准备 …………………………… 103
4.3.1　知识点 1：ListView …………… 103
4.3.2　知识点 2：对话框（Dialog） … 106
4.3.3　知识点 3：进度条（ProgressBar） …………………… 112
4.3.4　知识点 4：菜单 ……………… 118
4.4　项目实施 …………………………… 122
4.5　技术拓展 …………………………… 128
4.6　本章小结 …………………………… 130
4.7　强化练习 …………………………… 130

第 5 章 Android 数据存储 ……………131

- 5.1 项目导引…………………………131
- 5.2 项目分析…………………………132
- 5.3 技术准备…………………………132
  - 5.3.1 知识点 1：文件存储………132
  - 5.3.2 知识点 2：SharedPreferences……138
  - 5.3.3 知识点 3：嵌入式数据库 SQLite…………………………142
- 5.4 项目实施…………………………145
- 5.5 技术拓展…………………………148
- 5.6 本章小结…………………………151
- 5.7 强化练习…………………………151

第 6 章 Android 多媒体………………152

- 6.1 项目导引…………………………152
- 6.2 项目分析…………………………152
- 6.3 技术准备…………………………153
  - 6.3.1 知识点 1：Android 网络基础（标准 Java、Apache、Android 网络和 HTTP 通信接口）……………………………153
  - 6.3.2 知识点 2：Service…………157
  - 6.3.3 知识点 3：MediaPlayer……159
  - 6.3.4 知识点 4：视频……………168
  - 6.3.5 知识点 5：录音……………170
- 6.4 项目实施…………………………175
- 6.5 技术拓展…………………………182
- 6.6 本章小结…………………………186
- 6.7 强化练习…………………………186

第 7 章 综合实训 1 手机乐游项目……187

- 7.1 项目分析…………………………187
- 7.2 项目设计…………………………189
- 7.3 项目实施…………………………192
- 7.4 项目成果…………………………196

第 8 章 综合实训 2 基于 Android 的手机定位项目………………………197

- 8.1 项目分析…………………………197
- 8.2 项目设计…………………………197
- 8.3 项目实施…………………………199
  - 8.3.1 我在哪儿……………………201
  - 8.3.2 电子地图……………………202
  - 8.3.3 历史记录……………………204
  - 8.3.4 周边搜索……………………223
  - 8.3.5 线路规划……………………228
  - 8.3.6 分享给好友…………………232
- 8.4 项目成果…………………………233

参考教材……………………………………234

第1章 Android 开发环境构建

1.1 项目导引

2007 年 11 月 5 日，这个平常却又值得纪念的日子，Google 联合摩托罗拉、华为、宏达电、三星、LG 等 33 家手机厂商，以及手机芯片提供商、软硬件供货商、移动运营商联合组成开放手机联盟（Open Handset Alliance，OHA）和 Android 操作系统开放平台。Android 的推出让全球的程序员团结起来，加入到了手机开发的行列。

伴随着第一个 Android SDK 版本（m3-rc20a）的诞生，Google 同时举办了一场 375 万美元的创意挑战赛，这场堪称有史以来奖金最丰厚的创意大赛，全球共有 1 800 支队伍参加，让 Android 在短短的 3 个月内成为全球程序员都知道的"机器人"。2008 年 9 月 22 日，经过 Android SDK 的几次改版，美国移动运营商 T-Mobile USA 与 HTC 正式推出第一款 Google 手机——G-phone，代号"G1"，也让 Android 从模拟正式成为机器人。到目前为止，Google 手机已经从当初的"G1"发展到了"G9"；2012 年 11 月底，Google 宣布推出新的 Android SDK 的版本——SDK 4.2。值得一提的是，Google 在声明中提到："Android 设计的初衷就是向下延伸到主流手机，并向上扩展到 MID（移动上网设备）及小型设备"，现在以 Android 为操作系统的平板电脑已经面世，相信在不久的将来，Android 的发展将会越来越广阔。

从本章开始我们就带领大家进入 Android 的世界。通过本章的学习，大家可掌握 Android 开发环境的构建，Android 应用程序框架以及 Android 生命周期等基本概念。

1.2 项目分析

作为一个 Android 应用程序开发人员，掌握 Android 开发环境的构建是必须的。只有掌握了最基本的开发环境，后续的学习和工作才能顺利开展。在本章，我们从开发环境的构建开始，然后完成第一个 Android 程序。重点掌握创建 Android 应用程序的步骤和注意事项。

1.3 技术准备

1.3.1 Android 系统架构

上面我们谈到了 Android 备受青睐，那么是什么样的原因让 Android 如此受欢迎呢？下面让我们更深入地了解一下 Android 的系统架构。从软件分层的角度来看，Android 平台由应用程序、应用程序框架、系统库、Android 运行时以及 Linux 内核 5 部分组成。

Android 的系统框架如图 1-1 所示。

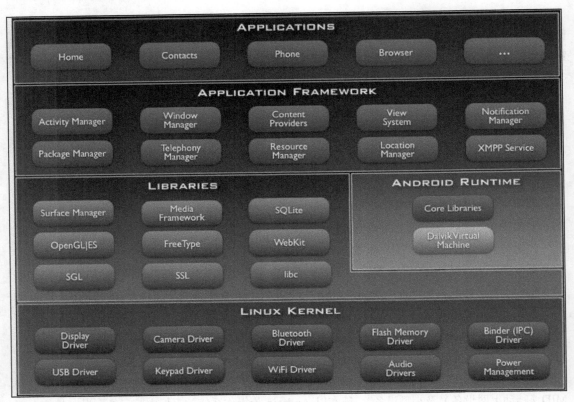

图 1-1　Android 系统架构

下面我们详细介绍 Android 平台的系统架构。

1. 应用程序

Android 平台默认包含了主要的应用程序，包括电子邮件、短信、日历、地图、浏览器、联系人等，这些程序都是用 Java 语言来编写的，当然你也可以用自己编写的软件来替代 Android 提供的程序。也许你会问为什么，请带着这个问题了解下面将要介绍的应用程序框架。

2. 应用程序框架

Android 应用程序框架是开发人员进行开发的基础。首先了解一下应用程序框架所包含的内容。它包括 9 大部分：视图系统、内容提供器、窗口管理器、活动管理器、通知管理器、位置管理器、资源管理器、电话管理器和包管理器。下面逐一介绍各部分的功能。

- 视图系统（View System）：用来构建应用程序的基本组件，包括文本框、按钮、列表等，甚至内嵌的网页浏览器。
- 内容提供器（Content Provider）：提供了程序之间数据的共享机制，例如我们可以在某个应用程序中调用本地数据库中的音频和视频文件等。
- 窗口管理器（Window Manager）：管理所有的窗口程序。通过窗口管理器提供的接口我们可以向窗口中添加 View，当然也可以同窗口中删除 View。
- 活动管理器（Activity Manager）：管理应用程序生命周期，并提供导航回退功能。
- 通知管理器（Notification Manager）：使所有的程序能够在状态栏显示自定义的警告。需要注意的是，在手机上的状态栏位于屏幕的顶部，例如，手机短信已满的提示就出现在状态栏。
- 位置管理器（Location Manager）：用来提供位置服务。其中包括两种技术：GPS 定位技术和网络定位技术。
- 资源管理器：（Resource Manager）：提供各种资源让应用程序使用，如布局文件、图片、音频文件等非代码资源。
- 电话管理器（Telephone Manager）：管理所有的移动设备。
- 包管理器（Package Manager）：主要用于系统内的程序管理。

在 Android 平台中，开发人员可以完全访问核心的应用程序所使用的 API，可以自由地利用设备硬件优势，访问位置信息、运行后台服务、设置闹钟、向状态栏添加通知等来开发出更多实用和新颖的程序。同时，Android 平台在设计时就考虑到了组件的重用，基于这种机制，用户就可以方便地替换平台本身所提供的各种应用组件。开发人员在开发 Android 平台上的应用程序时，也可以使用新的软件组件，并将该软件组建放入 Android 的应用程序框架中。

3．系统库

应用程序框架是贴近于应用程序的软件组建服务，而更底层则是 Android 的函数库。其架构如图 1-2 所示。

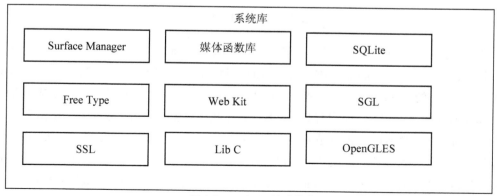

图 1-2　Android 的系统库

系统库的各个部分的功能如下。

- Surface Manager：在同时执行多个应用程序时，Surface Manager 会负责管理显示与存取操作之间的互动，并且为应用程序提供 2D 和 3D 图层的无缝融合。
- 媒体函数库：基于 OpenCORE（PacketVideo）；该库支持录放，并且可以录制许多流行的音频、视频格式，还有静态影像文件，包括 MPEG4、MP3、AAC、AMR、JPG、PNG。
- SQLite：一个对于所有应用程序可用的，轻量级的关系型数据库引擎。
- Free Type：提供点阵字和向量字的描绘显示。
- Web Kit：一个最新的 Web 浏览器引擎，用来支持 Android 浏览器和一个可嵌入的 Web 视图。Web Kit 是一个开源的项目，许多浏览器也都是用 Web Kit 引擎开发而成的，如诺基亚 S60 手机内的浏览器。
- SGL：提供 Android 在 2D 绘图方面的绘图引擎。
- SSL：媒体框架，提供了对各种音频、视频的支持。Android 支持多种音频、视频、静态图像格式等，如 MPEG-4、AMR、JPG、PNG、GIF 等。
- Lib C：一个从 BSD 继承来的标准 C 系统函数库，专门为基于嵌入式 Linux 的设备定制。
- OpenGLES：提供 OpenGLES 1.0 APIs 实现；该库可以使用硬件 3D 加速（如果可用）或者使用高度优化的 3D 软加速。

4．Android 运行时

Android 虽然采用 Java 语言来编写应用程序，但是它并不使用 J2ME 来执行 Java 程序，而是采用 Android 自用的 Android 运行时。Android 运行时包括核心库和 Dalvik 虚拟机两部分，如图 1-3 所示，下面我们就介绍一下这两部分的功能。

图 1-3　Android 运行时

核心库：核心库包含两部分内容，一部分为绝大多数 Java 程序语言所需要调用的功能函数，如 java.io 等；另一部分是 Android 的核心库，如 android.os、android.net 等。注意：每个 Android 应用程序都有一个自有的进程，Android 不是用一个 Dalvik 虚拟机来同时执行多个 Android 应用程序，而是每个 Android 应用程序都用一个自有 Dalvik 虚拟机来执行，这点与标准的 Java 有所不同。

Dalvik 虚拟机：Dalvik 虚拟机是 Google 公司自己设计的用于 Android 平台的 Java 虚拟机。它是专门为移动设备而设计的，在开发的时候就考虑到了用最少的内存资源来执行。在设计时，Dalvik 虚拟机很多地方参考了 Java 虚拟机的设计，但是它并不支持 Java 虚拟机所执行的 Java 字节码，也不直接执行 Java 的类文件。它可以支持已转换为.dex（即 Dalvik Executable）格式的 Java 应用程序的运行，.dex 格式是专为 Dalvik 设计的一种压缩格式，适合内存和处理器速度有限的系统。（dx 是一套工具，可以将 Java .class 转换成 .dex 格式．一个 dex 档通常会有多个.class。由于 dex 有时必须进行最佳化，会使档案大小增加 1～4 倍，以 ODEX 结尾。）

注意：Dalvik 虚拟机与 Java 虚拟机的最大不同在于 Java 虚拟机是基于栈（stack-based），而 Dalvik 基于寄存器（register-based）。基于寄存器的虚拟机的其中一个优点是所需要的资源相对较少，在硬件实现上也会比较容易。

5．Linux 内核

Android 平台中的操作系统采用了 Linux 2.6 版本的内核，它包括了显示驱动、摄像头驱动、Flash 内存驱动、Binder（IPC：系统进程间通信）驱动、键盘驱动、WiFi 驱动、Audio 驱动以及电源管理部分。它作为硬件和软件应用之间的硬件抽象层（HAL，Hardware Abstraction Layer），使得应用程序开发人员不需关心硬件细节。但是对于硬件开发商而言，如果想使 Android 平台运行到自己的硬件平台上就必须对 Linux 内核进行修改，为自己的硬件编写驱动程序。

1.3.2 开发环境搭建

在开始之前，我们需要准备一下操作系统环境以及程序，如表 1-1 所示。

表 1-1　　　　　　　　　　　　　　操作系统环境及程序

支持的操作系统	支持的开发环境
Windows XP、Vista 或 Win7 Mac OS X 10.4.8 或更高版本（硬件必须是 X86 的版本） Linux（在 Ubuntu Dapper Drake 上测试通过）	Eclipse IDE： ➢ Eclipse 3.3/3.4 或更高版本 ➢ JDK 5 或 JDK 6（仅 JRE 不能满足要求） ➢ SDK 2.1 以上版本

注：本书的例子开发环境为 Eclipse 3.4 GANYMEDE 版本，Android 版本为 4.2，操作系统为 Windows XP

从表 1-1 中我们可以知道，Android 平台的开发同时支持 Windows、Linux 以及 Mac OS。对于软件环境，Android 支持两种情形下的设置，一种是采用 Eclipse + ADT（Android Development Tools 插件），一种是自己构建开发环境。在这里我们推荐使用前者。

在进入下节之前，请先准备如下软件：

Eclipse 3.3/3.4 下载地址：http://www.eclipse.org/downloads/

Android SDK 4.2 下载地址：http://developer.android.com/sdk/index.html，可以选择不同的开发平台进行下载，如图 1-4 所示。

Platform	Package	Size	MD5 Checksum
Windows 32 & 64-bit	android-sdk_r21.1-windows.zip	99360755 bytes	dbece8859da9b66a1e8e7cd47b1e647e
	installer_r21.1-windows.exe (Recommended)	77767013 bytes	594d8ff8e349db9e783a5f2229561353
Mac OS X 32 & 64-bit	android-sdk_r21.1-macosx.zip	66077080 bytes	49903cf79e1f8e3fde54a95bd3666385
Linux 32 & 64-bit	android-sdk_r21.1-linux.tgz	91617112 bytes	3369a439240cf3dbe165d6b4173900a8

图 1-4　Android SDK 支持平台

JDK 5以上版本 网址：http://www.oracle.com/technetwork/java/javase/downloads/index.html

1. 安装 JDK

在 Windows 上安装 JDK 比较简单，下载 JDK 的安装文件后单击.exe 安装程序，按照提示，一直单击"下一步"按钮就可完成安装。在完成安装后还需要进行设置。配置步骤如下。

（1）右键单击"我的电脑"，打开属性页面。然后选择"高级"选项卡里面的"环境变量"，如图 1-5 所示。

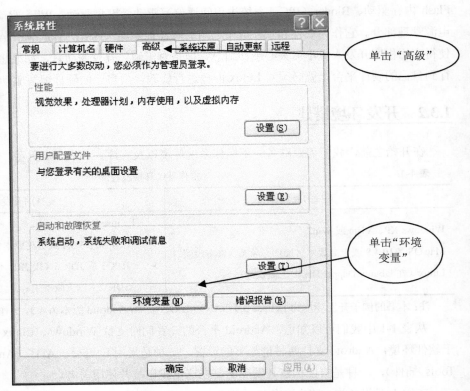

图 1-5 系统环境变量

（2）在新打开界面中的系统变量需要设置 3 个属性"java_home"、"path"、"classpath"，在没安装过 jdk 的环境下，path 属性已存在的，而 java_home 和 classpath 是不存在的。由于作者的计算机上已经安装并配置了 JDK，所以能够看到 classpath 和 java_home 选项，如图 1-6 所示。

（3）接下来就要配置各个选项。在没有配置过 JDK 的情况下，首先单击"新建"按钮，变量名为"java_home"，变量值为刚才 JDK 的安装路径，由于作者安装的路径为"D:\Program Files\Java\jdk1.6.0_07"，所以 java_home 的配置结果如图 1-7 所示。

（4）配置完"java_home"之后，下面来配置"path"，在系统变量里找到 path，然后单击"编辑"按钮，path 变量的含义就是系统在任何路径下都可以识别 java 命令，其变量值为"%java_home%\bin"（其中"%java_home%"的意思为刚才设置 java_home 的值）。

注意：在配置 path 的时候只需要将变量值直接加到原来的变量值后面即可，在添加的时候，记住在%java_home%\bin 之前添加一个分号，即"；%java_home%\bin"。

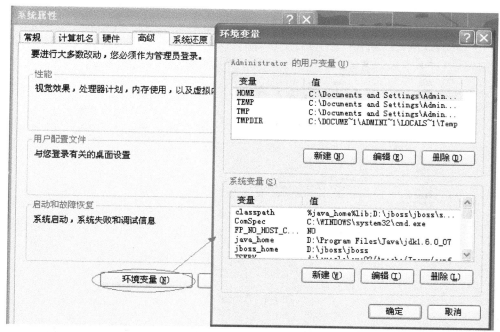

图 1-6　环境变量选项

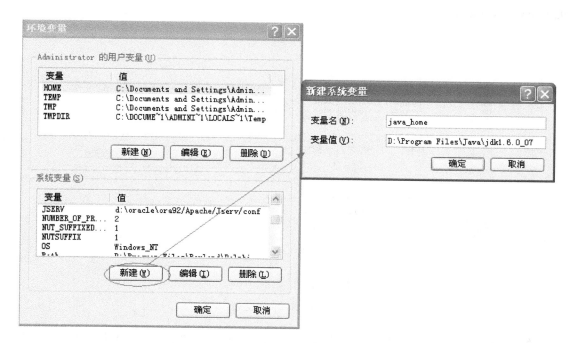

图 1-7　java_home 配置

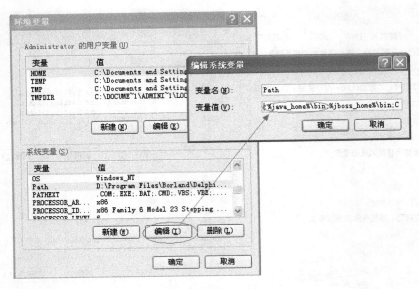

图 1-8　Path 配置

（5）最后，配置"classpath"，同样首先单击"新建"，然后在变量名上写 classpath，该变量的含义是为 java 加载类的路径，只有类在 classpath 中，java 命令才能识别，其值为".;%java_home%\lib;%java_home%\lib\tools.jar"。注意：在变量值前面要加"."表示当前路径。classpath 配置如图 1-9 所示。

图 1-9　classpath 配置

（6）以上 3 个变量设置完毕，则按"确定"按钮直至属性窗口消失，下面来验证看看安装是否成功。先打开"开始"→"运行"，输入"cmd"，进入 DOS 系统界面。然后输入"java –version"，如果安装成功。系统会显示所安装的 JDK 的版本信息（版本号不同则显示不同），如图 1-10 所示。

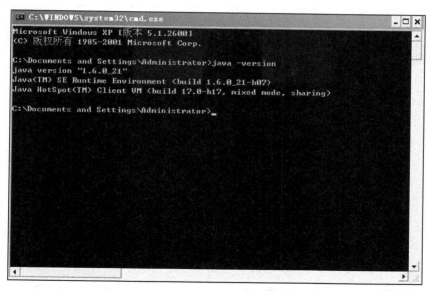

图 1-10　测试 jdk

2．安装 SDK

首先需要将下载下来的 SDK 压缩文件解压到一个目录中，这里我们将 SDK 安装到 "D:\android-sdk_r06-windows\android-sdk-windows" 中，如图 1-11 所示。

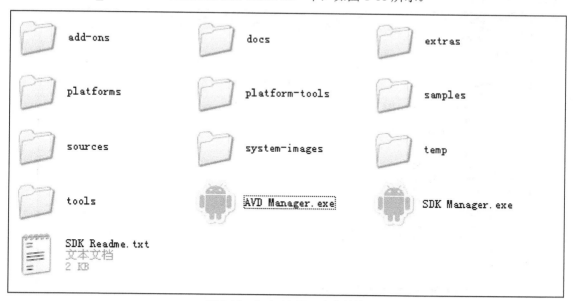

图 1-11　Android SDK 安装目录

双击 "SDK Setup.exe" 文件，运行 "Android SDK and AVD Manager" package 下载程序。注意：在双击 "SDK Setup.exe" 之前如已经安装 Eclipse 且正在运行，请先关闭 Eclipse 开发程序。若程序运行后并没有在 "Avaliable Package" 里发现 Android SDK Packages，那么请先切换到 "Settings" 选项，然后勾选 Misc 选项里的 "Force https://...source to be fetched using http://..." 如图 1-12 所示。

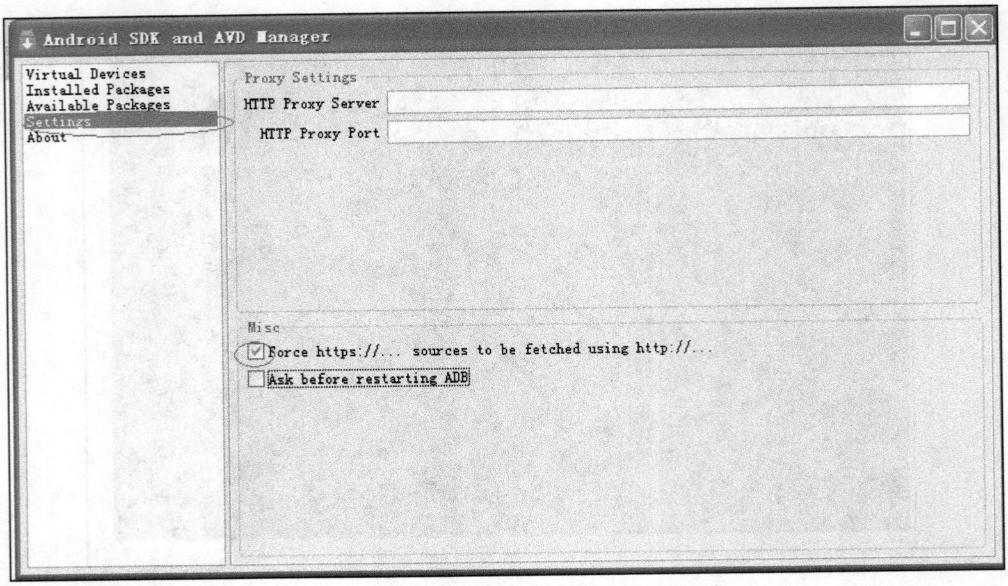

Android 应用程序设计

图 1-12　指定以 http://协议获取 package 列表

然后切换到"Avaliable Packages"选项中，勾选所要安装的 SDK Packages 或 API，按下"Installed Packages"按钮继续安装，若在"Avaliable Packages"列表中仍没有显示 SDK Source 则返回 Packages 列表，单击"Refresh"按钮刷新，如图 1-13 所示。

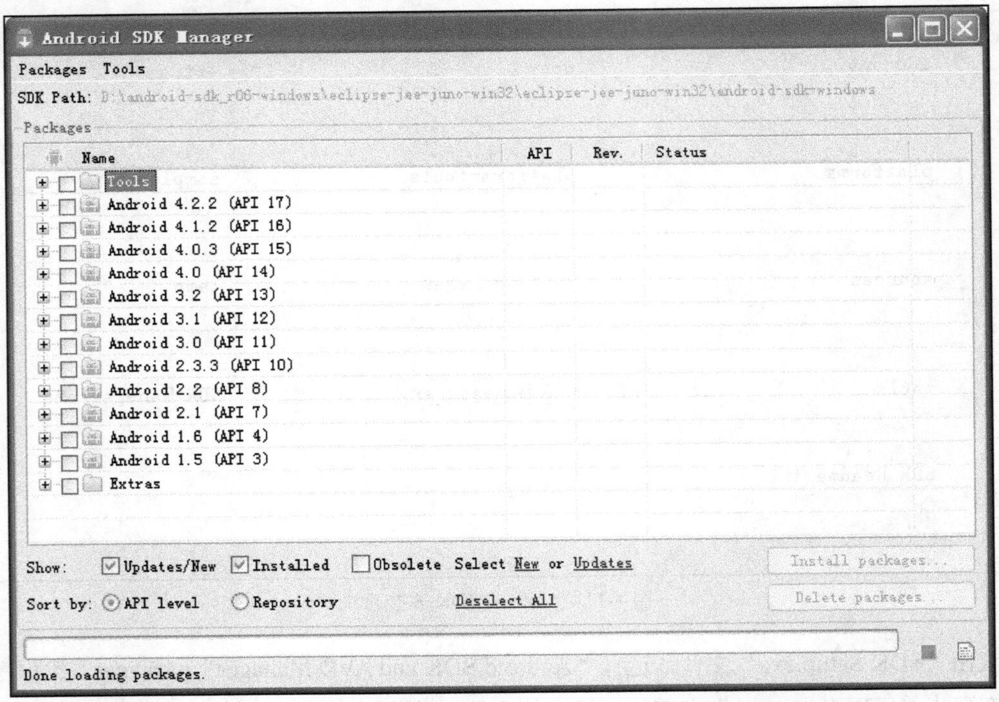

图 1-13　选择所要安装的 Packages，单击"Install packages"进行安装

单击"Install packages"后，会进入安装界面，选择"Accept License"然后单击"Install"按钮进行安装，如图 1-14 所示。

10

第 1 章 Android 开发环境构建

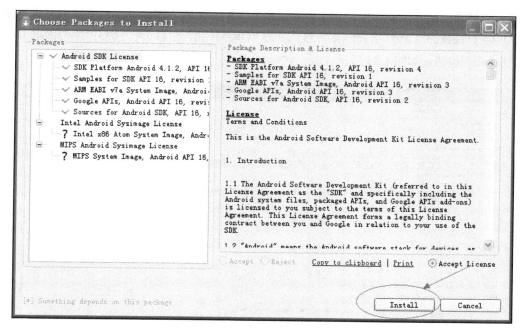

图 1-14 单击 "Install" 开始下载 SDK

至此，Android SDK 就安装完成了。

3. 安装 Eclipse 和 ADT

Eclipse 的安装不需要执行安装程序，直接将下载下来的压缩文件解压到一个指定的文件夹下即可，不过请注意，在此之前请确保已经正确地安装了 JDK。本书采用的 Eclipse 版本为 Eclipse 3.4 GANYMEDE 版本。在解压好 Eclipse 后安装 ADT(Android Development Tools)Plug-in。

（1）启动 Eclipse，选择 "Help" → "Software Updates"，如图 1-15 所示。

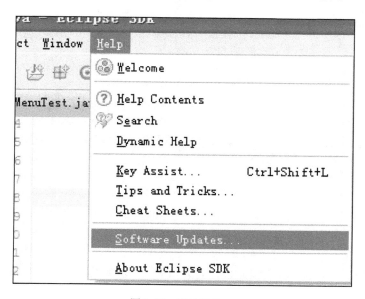

图 1-15 启动 Eclipse

11

（2）选择 Available Software 选项卡，然后单击"Add Site"按钮，在 Add Site 对话框中输入 https://dl-ssl.google.com/android/eclipse/，然后单击"OK"按钮，在对话框中选中刚才添加的网址，单击"Install"按钮，如图 1-16 所示。

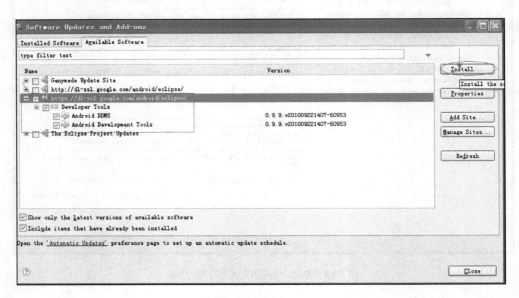

图 1-16　找到 ADT

（3）直接单击"Finish"按钮来安装所选择的 ADT，如图 1-17 所示。

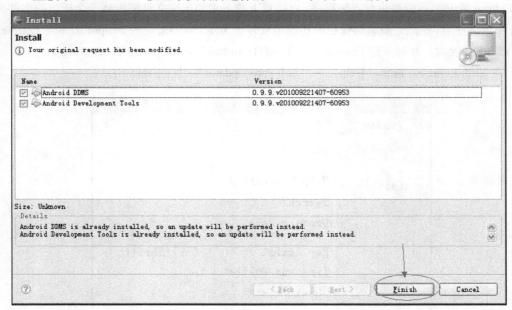

图 1-17　确认安装

（4）ADT 的安装过程可以放到后台运行，如图 1-18 所示，单击"Run in Background"按钮。

第 1 章 Android 开发环境构建

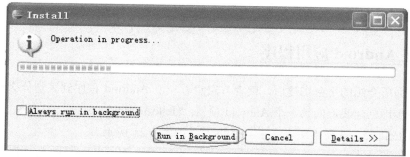

图 1-18　ADT 安装

（5）ADT 安装完成后需要重新启动 Eclipse，在 Eclipse 重启后要进行 Android ADT 插件的设置。选择"Window"→"Preferences"，如图 1-19 所示。

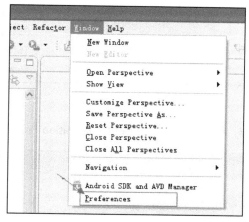

图 1-19　打开 Preference 对话框

（6）单击"Android"，然后单击"Browse"按钮，找到之前 Android 安装路径，单击"确定"。

注意：在 Eclipse 中 SDK 的路径配置中，SDK Location 的路径应该为 Android SDK and AVD Manager Setup 的路径，而不是 android-sdk_r06-windows 的路径，故在本书的配置例子中应该设置为 D:\android-sdk_r06-windows\android-sdk-windows，如图 1-20 所示。

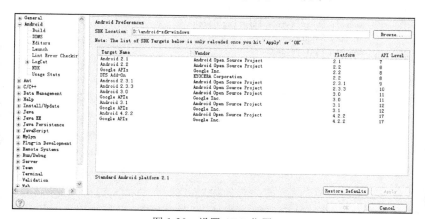

图 1-20　设置 ADT 位置

13

至此，Android 开发环境的搭建都已经完成。

1.3.3　创建 Android 应用程序

在完成前面所介绍的平台搭建后，读者肯定想看一下 Android 程序到底是什么样子的。那么下面就来一步步地创建我们的第一个 Android 项目：HelloAndroid。

（1）打开 Eclipse，选择"New"→"Project"，如图 1-21 所示。

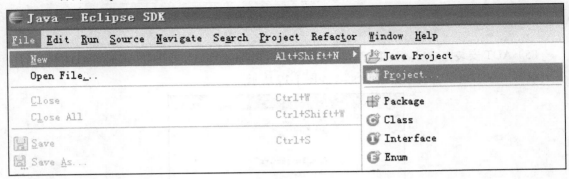

图 1-21　新建项目选项

（2）在"New Project"对话框中选择"Android Application Project"选项，如图 1-22 所示。

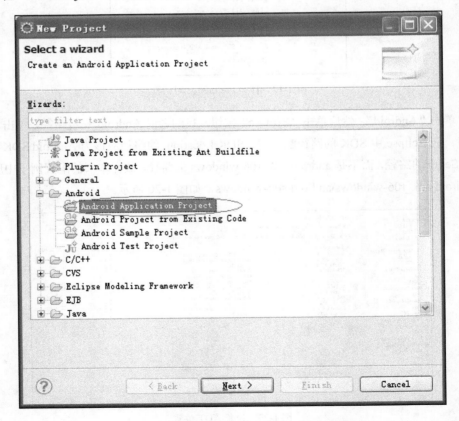

图 1-22　选择新建项目类型 Android Application Project

（3）单击"Next"按钮进入下一步，需要填写项目的名称等属性，如图 1-23 所示。

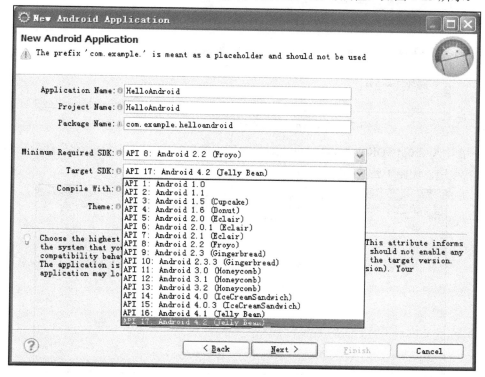

图 1-23 填写项目属性

其中：

- Project Name 表示项目的名称；
- Application Name 表示应用程序的名称，即显示在应用程序上的标题栏；
- Package Name 为包的名称，Java 中采用包名称来区分不同的类，设为 sziit.practice.chapter1；
- Create Activity 默认为选择此项，表示 ADT 插件将会创建一个默认的类，这个类是 Activity 类的子类。Activity 类是一个用来启动和控制程序的类；
- Minimum Required SDK 这个值对应了应用程序所要求的 API 的版本，由于不同设备上的 Android 系统映像各有不同，我们通过指定应用程序所要求的 API 版本将有助于应用程序的管理。如果设备上的系统映像的 API 版本低于应用程序的要求，那么应用程序将无法安装。在这里我们写的是 8，也就是对应的 SDK 版本为 2.2。

（4）然后单击"Finish"按钮，至此，你的第一个 Android 应用程序已经创建好了（也许在新的工作空间中，第一次创建完项目后会在项目的文件夹下出现红色的"x"说明项目有错误，不用担心，只要重新启动改一下 Eclipse 就可以了），对于这个新建的 Android 项目，即使没有书写任何一行代码，就已经可以运行。但是请注意：这个项目需要在 Android 模拟器（Emulator）上来运行。首先需要创建一个 Android Virtual Devices。创建模拟器很简单，通过 Android SDK and AVD Manager 可以很轻松地完成对模拟器的创建，如图 1-24 所示。

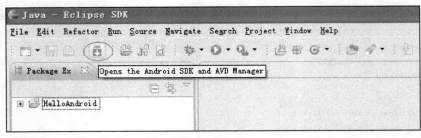

图 1-24 单击 Opens the Android SDK and AVD Manager

（5）在"Android SDK and AVD Manager"对话框中选择 "Virtual Devices"选项，然后选择"New…"按钮，如图 1-25 所示。

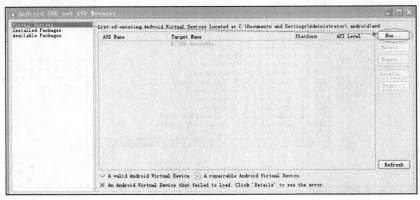

图 1-25 选择"New"按钮

（6）在弹出的"Edit Android Virtual Deuice（AVD）"对话框中填入创建的模拟器的名称以及 SD 卡的大小或者文件路径、屏幕默认分辨率等，我们可以选择模拟器的设备类型，如三星的 Galaxy 系列、Google Nexus 系列等。这里需要注意，在计算机上新建模拟器时需要调整默认的手机的 RAM 数值，由原来的 1024 改为 500 以下。如图 1-26 所示。

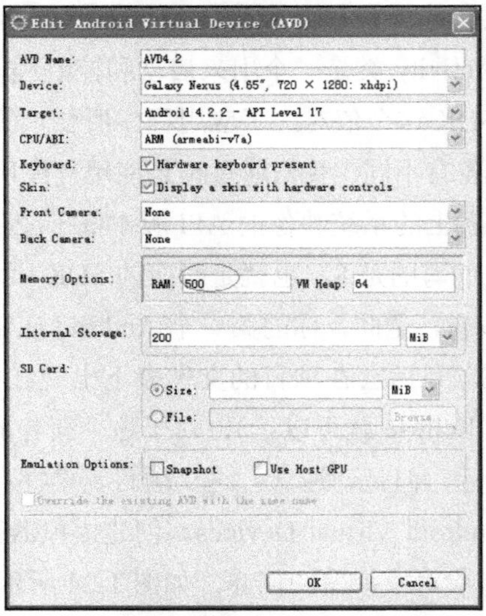

图 1-26 输入模拟器的名称等信息

（7）最后单击"OK"按钮，虚拟设备就创建好了。新建的模拟器就会在 Android SDK and AVD Manager 的 Virtual Devices 列表里显示出来。现在我们可以先让模拟器启动起来，如图 1-27 所示，单击"Start"按钮，然后选择"Launch"来启动刚才所创建的 AVD4.2.2 这个模拟器。注意：因为模拟器的启动需要一定的时间，我们在平常的学习中只需要启动一次就可以了，没有必要每次运行程序的时候都要重启模拟器。这样就可以省去等待模拟器启动的时间了。

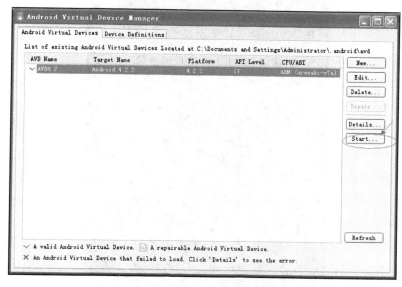

图 1-27　启动模拟器

（8）下面就可以运行刚才所创建的"HelloAndroid"项目了。在"Package Explorer"窗口中，单击刚创建的项目文件夹，在项目上右击，选择"Run As"→"Android Application"，如图 1-28 所示。

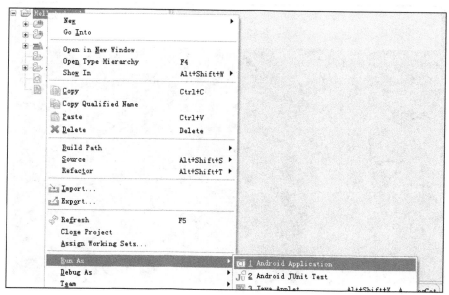

图 1-28　运行 HelloAndroid

这样就可以运行所新建的项目了，运行画面也如同在真的手机上一样，程序显示效果如图 1-29 所示。

图 1-29 运行效果

我们还可以按下"Ctrl+F12"组合键来切换布局，切换后运行效果如图 1-30 所示。

图 1-30 按下"Ctrl+F12"组合键后效果图

第 1 章 Android 开发环境构建

如果要退出程序的话，可以单击模拟器上的退格键，如图 1-31 所示。

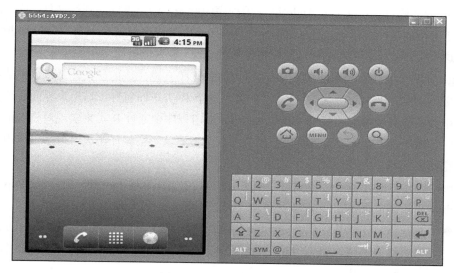

图 1-31　按退格键退出程序

虽然我们在程序中没有写任何代码，但是向导已经帮我们写好了程序的进入点、布局文件、字符串常数、应用程序访问权限等，除了作为应用模板之外，这也是 Android 手机程序的最佳学习范例。

1.3.4　解析 Android 应用程序框架

现在你也许会有疑问，我什么都没做为什么程序就能运行了呢？我们上文中说的向导到底是什么呢？它帮我们做了哪些工作就使得程序可以运行了呢？带着这些问题我们将对 Android 程序框架进行详细地分析。

1．应用程序结构

首先让我们看一下应用程序的构成。展开"Package Explorer"窗口中的"HelloAndroid.java"项目名称，可以看到如图 1-32 所示的目录结构。

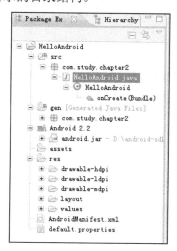

图 1-32　应用程序目录结构

19

目录中的 src、gen、res 目录和 AndroidManifest.xml 文件详细介绍如下。

- src 源代码目录：包含了 Android 应用程序中所需要的全部的源代码文件。这些源文件位于对应的包的目录下。在我们第一个 Android 应用程序中只有一个"HelloAndroid.java"，这个文件就是我们在创建项目时所输入的 Activity Name。
- gen 目录：gen 目录下面的 R.java 文件是由 ADT 自动生成的，文件是只读模式的，不能更改。实际上 R.java 定义了一个 R 类，这个 R 类里面包含了与用户界面、图像、字符串等各种资源对应的编号（id）。Android 应用程序中通过 R 类来实现对资源的引用。同时，编译器也会查看资源列表，没有用到的资源就不会编译进去，为手机应用程序节省空间。
- res 目录：在 res 目录下又包含了 5 个子目录，其中有 3 个是 drawable 的目录：drawable-hdpi; drawable-ldpi; drawable-mdpi 另外两个目录是 layout 和 values。
    - ◆ drawable 目录用来存放.png、jpg 等位图文件，我们可以通过 Resource.getDrawable(id) 来获得该资源。hdpi、ldpi、mdpi 分别对应的是高分辨率图片、低分辨率图片和中等分辨率图片。
    - ◆ Layout 目录下是程序的布局文件，是.xml 形式的布局文件。在 android 应用程序中可以使用 xml 布局文件来描述应用程序的布局，同样我们可以在程序中直接通过 Java 来创建用户界面。使用 xml 文件的好处是简单、结构清晰、维护容易，但是也有缺点，那就是不能动态地控制更改程序的界面。
    - ◆ values 目录下包含了所有使用 xml 格式的参数描述文件，我们可以自己添加所需要的资源，如字符串（string.xml）、颜色（color.xml）、样式（style.xml）等。
- AndroidManifest.xml：系统的控制文件，它的作用就是告诉系统如何处理所创建的所有顶层组件（Activity、Service、IntentReceiver 以及 ContentProvider）。它是每个应用程序所必需的文件，位于应用程序根目录下面，描述了程序包的全局变量，包括公开的应用程序组件（Activity、Service 等）和每个组件的实现类。

2．代码分析

上面我们从结构上对一个 Android 项目中所包含的主要的目录结构做了简单的介绍，下面我们从程序的角度来进行分析。

（1）AndroidManifest.xml。

在"Eclipse"中双击"AndroidManifest.xml"，选择 AndroidManifest.xml 选项卡，如图 1-33 所示。

第 1 章 Android 开发环境构建

图 1-33 打开 AndroidManifest.xml

打开 AndroidManifest.xml，我们会看到如代码清单 1-1 所示的代码。

代码清单 1-1 HelloAndroid/AndroidManifest.xml
<?xml version=*"1.0"* encoding=*"utf-8"*?>
<manifest xmlns:android=*"http://schemas.android.com/apk/res/android"*
　　　　package=*" sziit.practice.chapter1"*
　　　　android:versionCode=*"1"*
　　　　android:versionName=*"1.0"*>
　　<application android:icon=*"@drawable/icon"* android:label=*"@string/app_name"*>
　　　　<activity android:name=*".HelloAndroid"*
　　　　　　　android:label=*"@string/app_name"*>
　　　　　　<intent-filter>
　　　　　　　　<action android:name=*"android.intent.action.MAIN"* />
　　　　　　　　<category android:name=*"android.intent.category.LAUNCHER"* />
　　　　　　</intent-filter>
　　　　</activity>
　　</application>
　　<uses-sdk android:minSdkVersion=*"8"* />
</manifest>

在代码清单 1-1 中，intent-filters 描述了 Activity 启动的位置和时间，每当一个 Activity 要执行一个操作时，它将创建出一个 Intent 的对象，这个 Intent 对象能够承载的信息可描述了你想要

做什么、想处理什么数据、数据的类型以及一些其他的信息。Android 将 Intent 对象中的信息与所有公开的 intent-filter 比较，找到一个最能恰当处理请求者要求的数据和动作的 activity。AndroidManifest.xml 中的其他标签的作用如表 1-2 所示。

表 1-2　　　　　　　　　　　　　　AndroidManifest.xml 分析

项	描　　述
Manifest	根节点，描述了 package 中所有的内容
Xmlns:android	包含命名空间的声明 Xmlns:android=http://schemas.android.com/apk/res/android，使得 android 中各种标准属性能在文件中使用，提供了大部分元素中的数据
Package	声明应用程序包
Application	包含 package 中 application 级别组件声明的根节点。此元素也可包含 application 的一些全局和默认的属性，如标签、icon、主题、必要的权限等。一个 manifest 能包含 0 个或者 1 个此元素（不能大于一个）
Android:icon	应用程序图标
Android:label	应用程序名字
Activity	用来与用户交互的主要工具。Activity 是用户打开一个应用程序的初始界面，大部分被使用到的其他页面也由不同的 Activity 所实现，并声明在另外的 Activity 标记中。注意：每一个 Activity 必须有一个<activity>标记与之对应，无论它给外部使用或者只用于自己的 package 中。如果一个 Activity 没有对应的标记，它将不能运行。另外，Activity 可以包含一个或多个<intent-filter>元素来描述它所支持的操作
Android:name	应用程序默认启动的 Activity，声明了指定一组组件支持的 Intent 值，从而形成了 intent-filter。除了能在此元素下指定不同类型的值，属性也能放到这里来描述一个操作所需的唯一的标签、icon 和其他信息
Intent-filter Action	组件支持的 Intent Action
Category	组件支持的 Intent Category
Uses-sdk	应用程序所使用的 SDK 版本

注意：在代码清单 1-1 <application android:icon="@drawable/icon">中的"@drawable/icon"表示了对 res/drawable 目录下的 icon.png 的引用。其他依此类推。

（2）strings.xml。

打开位于 res/values 目录下的 strings.xml，在文件中定义了程序中所用到的一些常量，如代码清单 1-2 所示。

代码清单 1-2 HelloAndroid/res/values/strings.xml

<?xml version="*1.0*" encoding="*utf-8*"?>
<resources>
　　<string name="*hello*">Hello World, HelloAndroid!</string>
　　<string name="*app_name*">HelloAndroid</string>
</resources>

这个文件很简单，只有两个标签，分别定义了字符串"hello"的值为"Hello World, HelloAndroid!"，

字符串"app_name"的值为"HelloAndroid"。这些资源的应用我们会在后续章节中详细介绍。

（3）布局文件 main.xml。

打开位于 res/layout 下的 main.xml，我们会看到如代码清单 1-3 中的源代码。

代码清单 1-3 HelloAndroid/res/layout/main.xml

<?xml version="1.0" encoding="utf-8"?>
<LinearLayout xmlns:android="http://schemas.android.com/apk/res/android"
　　android:orientation="vertical"
　　android:layout_width="fill_parent"
　　android:layout_height="fill_parent" >
<TextView
　　android:layout_width="fill_parent"
　　android:layout_height="wrap_content"
　　android:text="@string/hello" />
</LinearLayout>

下面我们来分析一下代码清单 1-3 中所包含的内容。

- <LinearLayout>：线性布局格式，在此标签中，所有的组件都是线性排列组成的。我们在后续章节中会详细介绍 Android 中内置的几种布局。
- android:orientation：用来确定 LinearLayout 的方向，值可以为 vertical 或者 horizontal。其中，vertical 表示从上到下垂直布局，horizontal 表示从左到右水平布局。
- android:layout_width 和 android:layout_height：用来指明在父控件中当前控件的宽和高，可以设定值，但是更常用的是"fill_parent"和"wrap_content"。其中，fill_parent 表示填满父控件，wrap_content 表示大小刚好足够显示当前控件里的内容。
- <TextView>标签定义了一个用来显示文本的控件，其属性值 layout_width 为填满整个屏幕，layout_height 则可以根据文字的大小进行更改。Android:text 定义了在文本框中所要显示的文字内容，这里引用了 strings.xml 中的 hello 所定义的字符串资源，即"Hello World, HelloAndroid!"也就是我们在程序运行时看到的字符串。

（4）R.java。

打开位于 gen 目录下的 R.java，如代码清单 1-4 中所示。

代码清单 1-4 HelloAndroid/gen/R.java

package sziit.practice. chapter1;
public final class R {
　　public static final class attr {
　　}
　　public static final class drawable {
　　　　public static final int icon=0x7f020000;
　　}
　　public static final class layout {

```
            public static final int main=0x7f030000;
        }
        public static final class string {
            public static final int app_name=0x7f040001;
            public static final int hello=0x7f040000;
        }
}
```

在代码清单 1-4 中我们可以看到这里定义了很多常量，也许你会发现它们的名字都与 res 文件夹下的文件名相同，这再次证明了 R.java 文件中所存储的是该项目所有资源的索引。有了这个文件，程序便可以很快地找到要使用的资源，由于这个文件是只读的，不能手动编辑，所以当项目中加入了新的资源时，只要刷新项目，R.java 文件便自动生成了所有资源的索引。

（5）HelloAndroid.java。

打开 src 下的 HelloAndroid.java，源代码如代码清单 1-5 所示。

代码清单 1-5 HelloAndroid/src/HelloAndroid.java

```
package sziit.practice.chapter1; //声明应用程序包名
import android.app.Activity; //引入Activity包
import android.os.Bundle; //引入Bundle包
public class HelloAndroid extends Activity {//从Activity类派生子类HelloAndroid
    public void onCreate(Bundle savedInstanceState) {//重写onCreate
        super.onCreate(savedInstanceState); //父对象调用onCreate
        setContentView(R.layout.main); //设置屏幕布局
    }
}
```

最后，我们分析主程序中的内容。第 1 行定义了 Java 包。第 2 行和第 3 行导入了程序中使用到的 Android Java 包。接下来的就是对 HelloAndroid Activity 的定义。主程序 HelloAndroid 类继承自 Activity 类。重写了 void onCreate()方法，onCreate()函数在 Activity 创建时就将被调用，也就是说程序启动时 onCreate()就将运行。在 onCreate()中只有两行代码，其中第一句是对基类的 onCreate()方法的调用，用来获取 Activity 的状态，第二句用 setContentView(R.layout.main)来设定 Activity 所要显示的布局文件，它是通过对 R 类的引用来实现的，实际上就是位于 res/layout/main.xml 文件。

1.4 项目实施

在开发环境构建过程中，我们需要注意以下几个方面。

（1）系统环境变量的配置：我们在做 Java 开发时首先需要对环境变量进行设置，在这里，我们不仅需要配置和安装 JRE，还需要安装和配置 JDK。

（2）SDK 安装：在 SDK 安装过程中，我们需要从 Google 服务器上进行下载和安装，有时候会出现连接问题，在这里，我们需要有足够的耐心。

(3) 创建第一个 Android 项目，当我们在创建第一个 Android 项目时会出现错误，项目上会有红色的叉号，这是系统的一个问题，并不是程序的问题，我们可以尝试通过单击"Project"→"Clean"来解决，如图 1-34 所示。

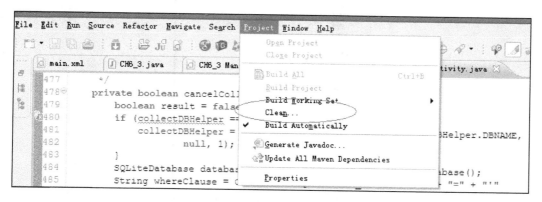

图 1-34　解决创建第 1 个 Andrord 项目时出现的错误

1.5　技术拓展：调试程序方法（DDMS 和 Logcat）

作为一名程序员，不可能保证程序都正确，所以经常遇到的问题就是调试程序。在 Eclipse 中调试程序的方法很多，使用 Eclipse 调试 Android 程序时需要注意细节。刚刚接触 Android 的开发者在调试 Android 程序的时候总是不能快速地找到程序的错误所在，因 Eclipse+ADT 的开发环境中没有直接跟踪对象内容的方法，但是我们可以用 ADT 插件中的 DDMS(Dalvik Debug Monitor Service)在 Eclipse 上轻松调试 Android 程序。DDMS 提供了很多功能，如测试设备截屏、Logcat、广播状态信息、模拟电话呼叫、接收 SMS、虚拟地理坐标等。下面我们通过 DDMS 来调试我们的第一个 Android 项目。

（1）首先，将 Eclipse 的工作界面切换到 DDMS 状态下。确定 Eclipse 开发工具右上角是否有 DDMS 标签，若有，可以直接单击该标签切换到 DDMS 工作界面，如图 1-35 所示。

图 1-35　切换到 DDMS 界面

若没有 DDMS 标签则需要打开"Open Perspective"按钮，然后选择"Other..."如图 1-36 所示。

图 1-36　打开 Open Perspective 界面

在弹出的"Open Perspective"对话框中选择"DDMS"选项，然后单击"OK"按钮，如图 1-37 所示。

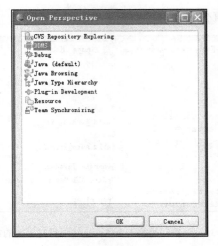

图 1-37 视图布局选择框

（2）在"DDMS"中选择"Devices"标签，可以查看其菜单的功能，其功能菜单如图 1-38 所示。其中包含 Debug Process（调试进程）、Update Heap（更新堆）、Cause GC（引起垃圾回收）、Update Threads（更新线程）、Stare Method Profiling（开始方法分析）、Stop Process（停止进程）、Screen Capture（屏幕截屏）、Reset adb（重启 Android Debug Bridge）菜单选项。

图 1-38 DDMS 功能菜单

通过 DDMS 中的菜单我们可以观察到 Android 程序运行时的各种状态，如进程信息、线程分析、堆内存的占用等。这些操作都是在 DDMS 框架下进行的，日常开发的程序是无法执行调用的。我们最常用的就是通过"Logcat"来调试 Android 程序。

在用"LogCat"来调试程序之前，先了解一下"LogCat"。它是通过"Android.util.Log"类的静态方法来查找错误和打印系统日志信息的，是一个进行日志输出的 API，我们在 Android 程序中可以随时为一个对象插入一个 Log，然后再观察"LogCat"的输出是不是正确。Android.util.Log 常用的方法有 5 个：

- Log.v(String tag,String msg); //Verbose, 冗余信息；
- Log.d(String tag,String msg); //Debug ,调试信息；
- Log.i(String tag,String msg); //Info, 普通信息；
- Log.w(String tag,String msg); //Warn 警告信息；
- Log.e(String tag,String msg); //Error 错误信息。

我们可以在程序中设置日志信息，然后运行程序，最后切换到"DDMS"界面来查看 LogCat 中输出的信息。更简单的方法是在"Java Perspective"界面下选择 "Window"→"Show View" →"Other…"，如图 1-39 所示。

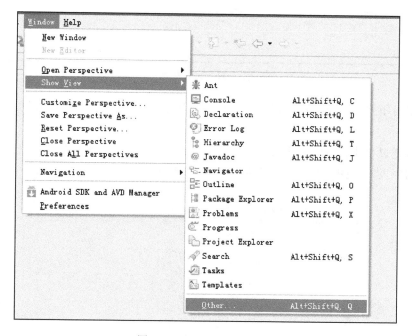

图 1-39　打开窗口显示菜单

在弹出的"Show View"对话框中选择"Android"目录下的"LogCat"然后单击"OK"按钮，就可以在 Java 工作界面的下方看到"LogCat"选项，如图 1-40 所示。这样就可以不用切换到"DDMS"界面下，从而看到日志的输出信息了。

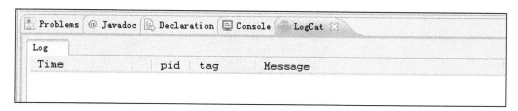

图 1-40　LogCat 视图窗口

另外，我们同样可以通过设置断点的方式来调试 Android 程序。在 Java 透视图中打开要设置断点的源文件，双击需要标记断点的代码前面的标记栏，就可以设置断点，如图 1-41 所示。

```
package sziit.practice.chapter1;//声明项目包名,指明本项目所有资源存放目录
import android.app.Activity;//导入活动(Activity)包
public class HelloAndroid extends Activity {//从活动(Activity)基类派生出子类HelloAndroid
    private static final String TAG= "HelloAndroid" ;
    /** Called when the activity is first created. */
    @Override
    public void onCreate(Bundle savedInstanceState) {//重写onCreate方法
        super.onCreate(savedInstanceState);//调用基类onCreate方法
        setContentView(R.layout.main);//设置屏幕布局
        Log.v(TAG, "onCreate");//撰写日志信息
    }
    public void onStart(){//重写onStart
        super.onStart();//调用基类onStart方法
        Log.v(TAG, "onStart");
    }
    public void onResume(){//重写onResume方法
        super.onResume();//调用基类onResume方法
        Log.v(TAG, "onResume");
    }
```

图 1-41　设置断点

注意：最好不要将多条语句放到同一行上，因为无法单步执行，也不能在同一行上为多条语句设置断点。

我们可以通过单步执行的方式来查找程序出错的位置，也可以声明条件断点，断点在表达式值发生变化时触发。如图 1-42 所示，我们设置条件"savedInstanceState==null"，当满足条件时，程序就会挂起。

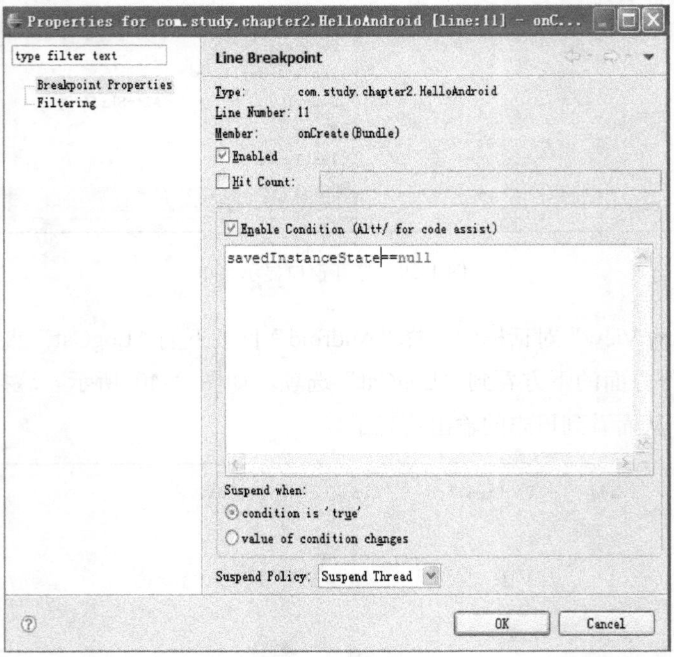

图 1-42　设置条件断点

当然还有很多的调试方法，上面只是举了两个简单的例子，读者可以根据自己的需要选择不同的调试方式，快速准确地找到程序的错误所在。

1.6 本章小结

在这一章中我们搭建了 Android 的开发平台，并创建了自己的第一个 Android 应用程序，对 Android 开发和调试有了初步的了解。至此，我想大家都知道为什么我们没有写一行代码，程序就能运行的原因了吧？请大家思考，Android 开发中布局文件用 xml 文件形式有什么好处呢？并且试着在我们创建的 HelloAndroid 项目上做修改，用输出日志的方式来调试第一个 Android 程序。

第一部分是 Android 程序开发的基础，希望读者能打好基础，接下来我们要对 Android 进行全面的学习。

1.7 强化练习

1. Android 系统平台框架一共分为哪几层？
2. Android 应用程序框架主要包含哪几部分？
3. Android 运行时的特点是什么？
4. Android 的内核版本是_____。
5. 在配置 Java 环境变量时我们需要配置的系统参数有哪些？
6. 在创建 Android 项目时，我们需要填写 min SDK Version 的意义是什么？

第 2 章 Android 屏幕布局

2.1 项目导引

随着 Android 操作系统的不断普及，越来越多的应用也随之诞生。怎么样才能设计和开发一款让用户接受和喜欢的 Android 应用已经越来越重要。那么这就需要我们了解 Android UI 设计。而谈到 UI，就不得不学习 Android 为我们提供的 Android 屏幕布局控件。

顾名思义，布局对象就是用于指明可视组件的布置方式，它本身是不可见的。我们希望通过本章的学习能够掌握 Android 屏幕布局的基本知识，并能够构建一个完美的屏幕布局。

2.2 项目分析

布局管理器（我们更习惯称之为布局）是 ViewGroup 的子类，用来控制子控件在屏幕中的位置。布局是可以嵌套的，因此可以使用多个布局管理器的组合来创建任何复杂的界面。在 Android SDK 中已经内置了几个简单的布局模型供我们使用，可以由用户决定选择哪些合适的布局组合来让界面更加利于理解和使用。下面我们将学习这几种布局模型，并配有实例以方便理解。

在 Android SDK 中主要包含以下几种布局。

➢ **LinearLayout**：线性布局，分为水平线性布局和垂直线性布局，是我们最常用的一种布局方式。在线性布局里面我们可以放多个控件，但是一行或者一列只能放一个控件。

➢ **RelativeLayout**：相对布局，与线性布局一样，在里面可以放多个控件，但是每个控件的位置都是相对的。我们可以定义每一个子 View 与其他子 View 之间以及与屏幕边界之间的相对位置。

➢ **TableLayout**：表格布局，在表格布局中我们可以使用多行多列的表格来布局 View。表可以跨越多行和多列，而且列可以设置为收缩或者增大的。

- **FrameLayout**：单帧布局，是最简单的布局管理器，它只是把控件放置在 View 的左上角，当我们添加一个新的 View 子类时，它会把每一个新的子 View 放到最上层。
- **AbsoluteLayout**：绝对布局或者坐标布局，顾名思义，在这个方式下的子 View 的位置都是绝对的。使用这个类的好处是我们可以使布局更加精确，但是却丧失了它的自适应的能力。
- **TabWidget**：切换卡，这是个特殊的布局模式，主要功能是实现标签切换，类似于 Android 系统"联系人"和"通话记录"的样式。

如何运用布局文件来完成项目需要呢？如何才能在项目中灵活运用布局呢？本章将首先学习布局的基本知识，然后通过项目实战来学习布局的使用。

2.3 技术准备

2.3.1 线性布局（LinearLayout）

线性布局是在开发中最常用到的布局方式之一，它提供了控件水平或者垂直排列的模型，但是需要注意在一行或者一列中只能放一个子 View。

LinearLayout 还支持为其包含的 widget 或者是 container 指定填充权值。好处就是允许其包含的 widget 或者是 container 可以填充屏幕上的剩余空间。这也避免了在一个大屏幕中，一串 widgets 或者是 containers 挤成一堆的情况，而是允许它们放大填充空白。剩余的空间会按这些 widgets 或者是 containers 指定的权值比例分配屏幕。默认的 weight 值为 0，表示按照 widgets 或者是 containers 实际大小来显示，若高于 0 的值，则将 container 剩余可用空间分割，分割大小具体取决于每一个 widget 或者是 container 的 layout_weight 及该权值在所有 widgets 或者是 containers 中的比例。例如，如果有 3 个文本框，其中两个指定的权值为 1，那么，这两个文本框将等比例地放大，并填满剩余的空间，而第 3 个文本框不会放大，按实际大小来显示。如果前两个文本框的取值一个为 2，一个为 1，则显示第 3 个文本框后剩余的空间的 2/3 给权值为 2 的，1/3 大小给权值为 1 的，也就是权值越大，重要度越大。

如果 LinearLayout 包含子 LinearLayout，子 LinearLayout 之间的权值越大的，重要度则越小。如果有 LinearLayout A 包含 LinearLayout C,D，C 的权值为 2，D 的权值为 1，则屏幕的 2/3 空间分给权值为 1 的 D，1/3 分给权值为 2 的 C。在 LinearLayout 嵌套的情况下，子 LinearLayout 必须要设置权值，否则默认的情况是未设置权值的子 LinearLayout 占据整个屏幕。

下面我们就通过一个简单的例子来学习线性布局的相关知识。在前面的例子中我们已经无数次用到了线性布局，现在就系统地进行学习。首先我们看一下这节演示实例 CH2_1 的运行截图，如图 2-1 所示。

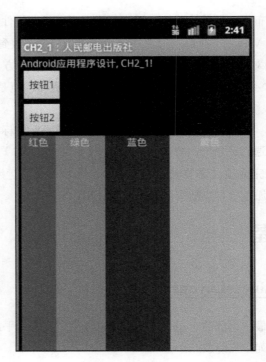

图 2-1　线性布局

下面就来分析一下它的布局文件的结构，源代码如代码清单 2-1 所示。

代码清单 2-1　第 2 章\CH2_1\res\layout\main.xml

<?xml version=*"1.0"* encoding=*"utf-8"*?>
<LinearLayout xmlns:android=*"http://schemas.android.com/apk/res/android"*
　　android:orientation=*"vertical"*
　　android:layout_width=*"fill_parent"*
　　android:layout_height=*"fill_parent"*
　　>
<TextView
　　android:layout_width=*"fill_parent"*
　　android:layout_height=*"wrap_content"*
　　android:text=*"@string/hello"*
　　/>
<Button
　　android:id=*"@+id/btnTest"*
　　android:layout_width=*"wrap_content"*
　　android:layout_height=*"wrap_content"*
　　android:text=*"按钮1"*/>
<Button
　　android:id=*"@+id/btnTest_1"*

android:layout_width=*"wrap_content"*
android:layout_height=*"wrap_content"*
android:text=*"按钮2"/>*
<LinearLayout xmlns:android=*"http://schemas.android.com/apk/res/android"*
 android:orientation=*"horizontal"*
 android:layout_width=*"fill_parent"*
 android:layout_height=*"fill_parent"*
 android:layout_weight=*"1"*>
<TextView
 android:text=*"红色"*
 android:gravity=*"center_horizontal"*
 android:background=*"#aa0000"*
 android:layout_width=*"wrap_content"*
 android:layout_height=*"fill_parent"*
 android:layout_weight=*"1"*/>
<TextView
 android:text=*"绿色"*
 android:gravity=*"center_horizontal"*
 android:background=*"#00aa00"*
 android:layout_width=*"wrap_content"*
 android:layout_height=*"fill_parent"*
 android:layout_weight=*"2"*/>
<TextView
 android:text=*"蓝色"*
 android:gravity=*"center_horizontal"*
 android:background=*"#0000aa"*
 android:layout_width=*"wrap_content"*
 android:layout_height=*"fill_parent"*
 android:layout_weight=*"3"*/>
<TextView
 android:text=*"黄色"*
 android:gravity=*"center_horizontal"*
 android:background=*"#aaaa00"*
 android:layout_width=*"wrap_content"*
 android:layout_height=*"fill_parent"*
 android:layout_weight=*"4"*/>
</LinearLayout>

</LinearLayout>

在本节的实例中我们首先采用了一个垂直方向的线性布局，宽度为占满整个屏幕（fill_parent），高度也为占满整个屏幕（fill_parent）。在此垂直的线性布局中有一个 TextView、两个按钮以及一个内嵌的线性布局。其中嵌入的线性布局方向为水平方向。在内嵌的线性布局中包括 4 个 TextView，它们分别为水平排列。下面我们就分析在布局中几个重要的属性，见表 2-1。

表 2-1　　　　　　　　　　　　　　线性布局及重要的属性

属　　性	描　　述
Android:orientation	确定 LinearLayout 的方向，取值可以为 vertical(垂直方向)或 horizontal(水平方向)
Android:layout_width Android:layout_height	指定当前控件在父控件中的宽和高，我们可以对其设定值，也可取值： Fill_parent:表示填满父控件的空白 Wrap_content:表示大小刚好足够显示当前控件的内容 Match_parent:它与 Fill_parent 的效果是一样的，从 SDK2.2 以后将 match_parent 来替代 fill_parent
Android:gravity	表示控件的对齐方式
Android:text	设置控件显示的名称

2.3.2　相对布局（RelativeLayout）

前面讲过的线性布局并不能够满足我们的需要，因为它本身存在着一个巨大的弱点，就是不能在一行或者一列中摆放多个控件，这也就增加了线性布局的局限性。在 Android 系统中提供了相对布局（RelativeLayout）的布局方式，顾名思义，相对布局的子控件会根据它们所设置的参照控件和参数进行相对布局，参照控件可以是父控件，也可以是其他的子控件。需要注意的是，不能在 RelativeLayout 容器本身和它的子元素之间产生循环依赖，例如，不能将 RelativeLayout 的高设置成为 WRAP_CONTENT 时将子元素的高设置为 ALIGN_PARENT_BOTTOM，因为如果容器是不定高的，那么子元素当然无法对齐容器的底边。

图 2-2 所示为我们通过一个相对布局进行排列的一个布局。

图 2-2　相对布局

下面先来看一下布局文件的源代码。

代码清单 2-2　第 2 章\CH2_2\res\layout\main.xml

```xml
<?xml version="1.0" encoding="utf-8"?>
<RelativeLayout
android:layout_width="fill_parent"
android:layout_height="fill_parent"
xmlns:android="http://schemas.android.com/apk/res/android"
>
<RadioButton
android:id="@+id/radioButton"
android:layout_width="wrap_content"
android:layout_height="wrap_content"
android:text="单选按钮(RadioButton) "
android:layout_alignParentTop="true"
android:layout_alignParentLeft="true"
>
</RadioButton>
<AnalogClock
android:id="@+id/clock"
android:layout_width="wrap_content"
android:layout_height="wrap_content"
android:layout_below="@+id/radioButton"
android:layout_alignParentLeft="true"
>
</AnalogClock>
<Button
android:id="@+id/btnButton"
android:layout_width="wrap_content"
android:layout_height="wrap_content"
android:text="按钮(Button) "
android:layout_alignTop="@+id/clock"
android:layout_alignParentRight="true"
>
</Button>
<TimePicker
android:id="@+id/timePicker"
android:layout_width="wrap_content"
```

```
android:layout_height="wrap_content"
android:layout_below="@+id/clock"
>
</TimePicker>
</RelativeLayout>
```

在相对布局中我们经常会用到一些常量，在表 2-2 中列出了几个常用的常量以及其相应的说明。

表 2-2　　　　　　　　　　　　　　　　常量及其说明

数据类型	常量名称	描述
int	ABOVE	定义将元素的底边对齐另一个元素的顶边
int	ALIGN_BASELINE	定义将元素基线的对齐另一个元素的基线
int	ALIGN_BOTTOM	定义将元素底边的对齐另一个元素的底边
int	ALIGN_LEFT	定义将元素的左边对齐另一个元素的左边
int	ALIGN_PARENT_BOTTOM	定义将元素的底边对齐其父容器(RelativeLayout)的底边
int	ALIGN_PARENT_LEFT	定义将元素的左边对齐其父容器(RelativeLayout)的左边
int	ALIGN_PARENT_RIGHT	定义将元素的右边对齐其父容器(RelativeLayout)的右边
int	ALIGN_PARENT_TOP	定义将元素的顶边对齐其父容器(RelativeLayout)的顶边
int	ALIGN_RIGHT	定义将元素的右边对齐另一个元素的右边
int	ALIGN_TOP	定义将元素的顶边对齐另一个元素的顶边
int	BELOW	定义将元素的顶边对齐另一个元素的底边
int	CENTER_HORIZONTAL	定义让元素在容器（RelativeLayout）内水平居中
int	CENTER_IN_PARENT	定义让元素位于容器（RelativeLayout）的中心
int	CENTER_VERTICAL	定义让元素在容器（RelativeLayout）内垂直居中
int	LEFT_OF	定义将元素的右边对齐另一个元素的左边
int	RIGHT_OF	定义将元素的左边对齐另一个元素的右边

此外，在 xml 的属性中有一个是需要注意的，那就是 android: ignoreGravity 属性，它表示设置容器中的哪个子元素会不受 Gravity 的影响。接收参数:子元素的 Id，如果设置成为 0，则全部子元素都会受到影响。除此之外我们再来看一下在 xml 中常用的属性。

第一组：设置的值为其他控件的 Id 号。

- android:layout_above：将该控件的底部置于给定 Id 的控件之上。
- android:layout_below：将该控件的底部置于给定 Id 的控件之下。
- android:layout_toLeftOf：将该控件的右边缘和给定 Id 的控件左边缘对齐。
- android:layout_toRightOf：将该控件的左边缘和给定 Id 的控件右边缘对齐。
- android:layout_alignBaseline：该控件的 baseline 和给定 Id 控件的 baseline 对齐。

- android:layout_alignBottom：将该控件的底部边缘与给定 Id 的底部边缘对齐。
- android:layout_alignLeft：将该控件的左边缘和给定 Id 的控件左边缘对齐。
- android:layout_alignRight：将该控件的右边缘和给定 Id 的控件右边缘对齐。
- android:layout_alignTop：将该控件的顶部边缘和给定 Id 的控件顶部边缘对齐。

第二组：设置的值为 Boolean 值，即 True 或者 False。

- android:alignParentBottom：如果值为 true，则将该控件的底部和父控件的底部对齐。
- android:alignParentLeft：如果值为 true，则将该控件的左边缘和父控件的左边缘对齐。
- android:alignParentRight：如果值为 true，则将该控件的右边缘和父控件的右边缘对齐。
- android:alignParentTop：如果值为 true，则将该控件的顶部和父控件的顶部对齐。
- android:layout_centerHorizontal：如果值为 true，该控件将被置于水平方向的中央。
- android:layout_centerInParent：如果值为 true，该控件将被置于父控件水平和垂直方向的中央。
- android:layout_centerVertical：如果值为 true，该控件将被置于垂直方向的中央。

2.3.3 表单布局（TableLayout）

表单布局（TableLayout）也是经常用到的一个布局方式，在表单布局中以行和列的形式来管理布局内的子控件，每一行为一个 TableRow 对象或者 View 对象。在 TableRow 对象中可以添加其他的子控件，不过要注意每添加一个子控件为一列。在表单布局中不会显示 Row、Column 或者 Cell 的边框线。与日常所用到的 HTML 的表单不一样的是，单元格不能跨列。我们可以在属性中设置列为隐藏的，或者是伸展的。

在表单布局中的子控件的宽度是要填满整个布局的，即属性：android:layout_width="fill_parent"。若子控件是 TableRow，则高度一定是刚好适合其内容，即属性：android:layout_height="wrap_content"，其他的控件则可以另外指定。TableRow 的子控件即每一个列可以指定其高度和宽度，但是必须为"wrap_content"或者"fill_parent"。在 SDK4.2.2 后我们可以用"match_parent"来替代"fill_parent"。下面就来通过实例 2_3 来对表单布局的应用做个直观的了解。在这个例子中我们通过程序来完成对表单布局的动态增加，如图 2-3 所示。

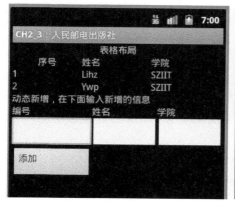
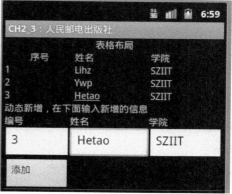

图 2-3　动态增加 TableLayout 的内容

程序布局文件的源代码如代码清单2-3所示。

代码清单2-3　第2章\CH2_3\res\layout\main.xml

```xml
<?xml version="1.0" encoding="utf-8"?>
<LinearLayout xmlns:android="http://schemas.android.com/apk/res/android"
    android:orientation="vertical"
    android:layout_width="fill_parent"
    android:layout_height="fill_parent">
<TableLayout xmlns:android="http://schemas.android.com/apk/res/android"
    android:id="@+id/tableLayout"
    android:layout_width="fill_parent"
    android:layout_height="wrap_content"
    android:stretchColumns="0,1,2"
    android:shrinkColumns="1,2"
    >
<TextView
    android:layout_width="fill_parent"
    android:layout_height="wrap_content"
    android:text="表格布局"
    android:gravity="center"
    />
<TableRow>
    <TextView
        android:text="序号"
        android:gravity="center"/>
    <TextView
        android:text="姓名"
        android:gravity="left"/>
    <TextView
        android:layout_width="wrap_content"
        android:layout_height="wrap_content"
        android:text="学院"
        android:gravity="left"/>
</TableRow>
<TableRow>
    <TextView
        android:text="1"
        android:gravity="left"/>
```

```xml
    <TextView
        android:text=" Lihz "
        android:gravity="left"/>
    <TextView
        android:text=" SZIIT "
        android:gravity="left"/>
</TableRow>
<TableRow>
    <TextView
        android:text="2"
        android:gravity="left"/>
    <TextView
        android:text=" "
        android:gravity="left"/>
    <TextView
        android:text=" SZIIT "
        android:gravity="left"/>
</TableRow>
</TableLayout>
<TableLayout
    android:layout_width="fill_parent"
    android:layout_height="wrap_content"
    android:stretchColumns="0,1,2"
    android:shrinkColumns="1,2"
    >
    <TextView
    android:layout_width="fill_parent"
    android:layout_height="wrap_content"
    android:text="动态新增，在下面输入新增的信息"
    android:gravity="left"
    />
<TableRow>
    <TextView
        android:text="序号"
        android:gravity="left"/>
    <TextView
        android:text="姓名"
```

```
            android:gravity="left"/>
        <TextView
            android:text="学院"
            android:gravity="left"/>
    </TableRow>
    <TableRow>
        <EditText
            android:id="@+id/Num"
            android:gravity="left"/>
        <EditText
            android:id="@+id/Name"
            android:gravity="left"/>
        <EditText
            android:id="@+id/College"
            android:gravity="left"/>
    </TableRow>
    <TableRow>
        <Button
            android:id="@+id/btnAdd"
            android:text="添加"
            android:gravity="left"/>
    </TableRow>
</TableLayout>
</LinearLayout>
```

在布局文件中我们在一个线性布局里嵌套了两个表单布局，在表单布局中又嵌套了 <TableRow> 对象。表 2-3 列举出了几个常用的属性以及对其进行的说明。

表 2-3　　　　　　　　　表单布局的 3 个重要属性及说明

属　　性	说　　明
Android:collapseColumns="1"	隐藏该 TableLayout 里的 TableRow 的列 1，注意编号是从 0 开始，当要隐藏多列时可以通过逗号隔开
Android:stretchColumns="1"	将列 1 设定为可伸展的列，它将尽量伸展以填满所有可用的空间。当有多个列需要设置时可以用逗号隔开
Android:shrinkColumns="1"	将列 1 设置为可收缩的列。当可收缩的列太宽以至于无法让其他列显示不全时则纵向延伸空间。多列设置时可以用逗号隔开

在主程序中需要完成动态添加信息的功能，为此，我们在布局文件中设置了一个按钮，在主

程序中只需要设置按钮的监听事件，在按钮的监听事件中完成对 TableLayout 布局的动态增加信息。程序代码如代码清单 2-4。

代码清单 2-4 第 2 章\CH2_3\src\com\study\chapter2\CH2_3.java

```java
public class CH2_3 extends Activity {//从基类Activity派生出CH2_3
    public void onCreate(Bundle savedInstanceState) {//派生类重载onCreate方法
        super.onCreate(savedInstanceState);  //基类调用onCreate方法
        setContentView(R.layout.main); //设置显示界面
        //声明各个对象并初始化，利用findViewById方法，通过Id索引找到相应的控件对象
        final TableLayout tablelayout=(TableLayout)findViewById(R.id.tableLayout); //表单布局
        Button buttonAdd=(Button)findViewById(R.id.btnAdd); //定义按钮（Button）对象
        final EditText edtNum=(EditText)findViewById(R.id.Num); //定义文本框（EditText）对象
        final EditText edtName=(EditText)findViewById(R.id.Name); //定义文本框（EditText）对象
        final EditText edtCollege=(EditText)findViewById(R.id.College); // 定义文本框对象
        final TableRow tableRow=new TableRow(this); //创建TableRow对象
        final TextView tvNum = new TextView(this); //创建文本框（EditText）对象
        final TextView tvName=new TextView(this); //创建文本框（EditText）对象
        final TextView tvCollege=new TextView(this); //创建文本框（EditText）对象
        //设置"添加"按钮的事件监听
        buttonAdd.setOnClickListener(new Button.OnClickListener(){
            public void onClick(View v) {//处理单击事件
                //设置TextView的取值
                tvNum.setText(edtNum.getText());
                tvName.setText(edtName.getText());
                tvCollege.setText(edtCollege.getText());
                //将设置的TextView添加到TableRow对象中
                tableRow.addView(tvNum);
                tableRow.addView(tvName);
                tableRow.addView(tvCollege);
                //将TableRow对象添加到TableLayout中
                tablelayout.addView(tableRow);
            }
        });
    }
}
```

在主程序中，只需要新建一个 TableRow 对象，将 TextView 对象添加到 TableRow 对象中，然后调用 TableLayout 对象的 addView()方法来将新建并赋值好的 TableRow 对象添加到 TableLayout 对象中。

2.3.4 单帧布局（FrameLayout）

单帧布局（FrameLayout）是最简单的一个布局模式。该布局 container 可以用来占有屏幕的某块区域来显示单一的对象，可以有多个 widgets 或者是 container，但是所有被包含的 widgets 或者是 container 必须被固定到屏幕的左上角，并且一层覆盖一层，不能为一个 widgets 或者是 container 指定一个具体位置。container 所包含的 widgets 或者是 container 的队列是采用堆栈的结构，最后加进来的 widgets 或者是 container 显示在最上面。所以后一个 widgets 或者是 container 将会直接覆盖在前一个 widgets 或者是 container 之上，把它们部份或全部挡住（除非后一个 widgets 或者是 container 是透明的，必须得到 FrameLayout Container 的允许）。我们先看一下本节实例的效果图，如图 2-4 所示。

图 2-4　单帧布局效果图

在布局文件中我们声明了 3 个 ImageView 对象，分别放入预先设计好的大、中、小 3 张图片，以此来增加显示的效果。布局文件的源代码如代码清单 2-5 所示。

代码清单 2-5　第 2 章\CH2_4\res\layout\main.xml

```
<?xml version="1.0" encoding="utf-8"?>
<FrameLayout xmlns:android="http://schemas.android.com/apk/res/android"
    android:orientation="vertical"
    android:layout_width="fill_parent"
    android:layout_height="fill_parent"
    >
<ImageView
    android:layout_width="wrap_content"
```

```
        android:layout_height="wrap_content"
        android:src="@drawable/big"/>
<ImageView
        android:layout_width="wrap_content"
        android:layout_height="wrap_content"
        android:src="@drawable/medium"/>
<ImageView
        android:layout_width="wrap_content"
        android:layout_height="wrap_content"
        android:src="@drawable/small"/>
</FrameLayout>
```

通过实例可以发现，所有图片都放到了左上角，而且最先放入的图片会被后面放入的图片覆盖掉。这就是单帧布局的效果，也可以看出，单帧布局属于最简单的一种布局方式。

2.3.5 坐标布局（AbsoluteLayout）

坐标布局（AbsoluteLayout）是指绝对布局，在布局上灵活性较大，也较复杂，另外由于各种手机屏幕尺寸的差异，给开发人员带来较多困难，所以它的使用比较少。

用坐标布局时，需要注意坐标原点为屏幕左上角，这和计算机屏幕的设置是一样的。添加视图时，要精确地计算每个视图的像素大小，最好先在纸上画草图，并将所有元素的像素定位计算好。我们可以使用可视化软件（DroidDraw）来进行布局设置。在设置绝对布局时，需要设置其属性：android:layout_x 和 android:layout_y 的值来确定子控件的位置。

在这个例子中我们通过 DroidDraw 对布局进行了设置，在 DroidDraw 中的效果如图 2-5 所示，布局完成后将布局文件放到工程中，运行效果如图 2-6 所示。

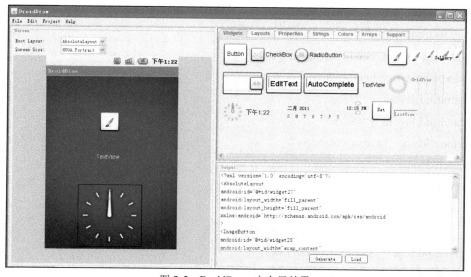

图 2-5　DroidDraw 中布局效果

图 2-6 实例中运行效果

布局文件的代码在软件中也会自动生成，如代码清单 2-6 所示。

代码清单 2-6 第 2 章\CH2_5\res\layout\main.xml

<?xml version=*"1.0"* encoding=*"utf-8"*?>
<AbsoluteLayout android:id=*"@+id/widget27"*
    android:layout_width=*"fill_parent"* android:layout_height=*"fill_parent"*
    xmlns:android=*"http://schemas.android.com/apk/res/android"*>
    <ImageButton android:id=*"@+id/imgButton"*
        android:layout_width=*"wrap_content"*
        android:layout_height=*"wrap_content"*
        android:layout_x=*"130px"*
        android:layout_y=*"92px"*
        android:src=*"@drawable/icon"*>
    </ImageButton>
    <TextView
        android:layout_width=*"wrap_content"*
        android:layout_height=*"wrap_content"*
        android:text=*"这是绝对布局的演示"*
        android:layout_x=*"120px"*

```
        android:layout_y="192px">
</TextView>
<AnalogClock android:id="@+id/analogClock"
        android:layout_width="wrap_content"
        android:layout_height="wrap_content"
        android:layout_x="80px"
        android:layout_y="272px">
</AnalogClock>
</AbsoluteLayout>
```

绝对布局中要求指定子控件的坐标，所以使其维护更加困难，缺乏灵活性，故使用范围很小。

2.3.6　切换卡（TabWidget）

切换卡（TabWidget）按照常理来说并不能算是一种布局方式，但是由于我们可以通过多个标签切换以显示不同的内容，所以暂且把它作为一种布局来讲解。首先来了解一下什么是 TabHost，它是一个用来存放多个 Tab 标签的容器，每个 Tab 都可以对应自己的布局。

要使用 TabHost，就需要通过 getTabHost 方法来获取 TabHost 对象，然后通过 addTab 方法来向 TabHost 中添加 Tab。需要注意，在每个 Tab 之间进行切换时都会产生一个事件，需要通过设置 setOnTabChangedListener()的事件监听捕获信息。

TabWidget 的使用有两种方法，一种是使用系统自带写好的 TabHost 及继承自 TabActivity 类；另一种是定义自己的 TabHost，不用继承 TabActivity。下面通过两个例子来看一下两者的区别。

（1）使用系统自带的 TabHost。此时我们需要注意：TabHost、TabWidget、FrameLayout 的 ID 必须分别为@android:id/tabhost、@android:id/tabs、@android:id/tabcontent。在此种情况下的 xml 布局文件的源代码如代码清单 2-7 所示。

代码清单 2-7　第 2 章\CH2_6\res\layout\main.xml

```
<?xml version="1.0" encoding="utf-8"?>
<TabHost
        android:id="@android:id/tabhost"
        android:layout_width="fill_parent"
        android:layout_height="fill_parent"
        xmlns:android="http://schemas.android.com/apk/res/android">
    <LinearLayout
            android:orientation="vertical"
            android:layout_width="fill_parent"
            android:layout_height="fill_parent">
        <TabWidget
            android:id="@android:id/tabs"
```

```
                android:layout_width="fill_parent"
                android:layout_height="wrap_content"/>
        <FrameLayout
            android:id="@android:id/tabcontent"
            android:layout_width="fill_parent"
            android:layout_height="fill_parent"
            android:padding="5dp">
            <TextView
                android:id="@+id/txt1"
                android:layout_width="fill_parent"
                android:layout_height="fill_parent"
                android:text="这是 切换卡1<tab>"/>
            <TextView
                android:id="@+id/txt2"
                android:layout_width="fill_parent"
                android:layout_height="fill_parent"
                android:text="这是 切换卡2<tab>"/>
        </FrameLayout>
    </LinearLayout>
</TabHost>
```

此时，在主程序中完成切换卡操作，代码如代码清单 2-8 所示。要注意 Activity 是继承自 TabActivity 的。

代码清单 2-8 第 2 章\CH2_6\src\com\study\chapter2\CH2_6.java

```java
//包的引入程序省略
public class CH2_6 extends TabActivity {//从切换卡基类TabActivity派生CH2_6类
    TabHost mTabHost;  //定义TabHost容器对象，用于存放多个Tab标签
    public void onCreate(Bundle savedInstanceState) {//派生类重载基类onCreate方法
        super.onCreate(savedInstanceState);  //基类调用onCreate方法
        setContentView(R.layout.main);  //设置屏幕布局，布局资源通过R.layout.main引用
        mTabHost=getTabHost();  //获取Tabhost对象
        //为Tabhost对象添加标Tab标签
        mTabHost.addTab(mTabHost.newTabSpec("tab1").setIndicator("TAB 1",getResources()
            .getDrawable(R.drawable.icon))  //设置切换卡第一个选项卡图标
            .setContent(R.id.txt1));  //设置切换卡第一个选项卡显示内容
        mTabHost.addTab(mTabHost.newTabSpec("tab2").setIndicator("TAB 2",getResources()
            .getDrawable(R.drawable.icon))  //设置切换卡第二个选项卡图标
            .setContent(R.id.txt2));  //设置切换卡第二个选项卡显示内容
```

mTabHost.setCurrentTab(0); //设置当前的Tab标签为第一个选项卡
//设置切换时的事件监听
mTabHost.setOnTabChangedListener(**new** OnTabChangeListener(){
    public void onTabChanged(String tabId) {//处理Tab切换事件响应
        Toast.*makeText*(CH2_6.**this**, "切换", Toast.*LENGTH_SHORT*).show();
    }
});
    }
}

在完成了布局文件和主程序后，运行下我们的程序，效果如图2-7所示。

图2-7 TabWidget 应用

（2）使用自己定义的 TabHost。此时，TabHost 的 ID 可以是自己定义的，但是 TabWidget、FrameLayout 标签的 ID 还是要求必须分别为@android:id/tabs、@android:id/tabcontent。此时的源代码如代码清单2-9所示。

代码清单 2-9 第 2 章\CH2_7\res\layout\main.xml

<?xml version=*"1.0"* encoding=*"utf-8"*?>
<TabHost
    android:id=*"@+id/tabhost_1"*
    android:layout_width=*"fill_parent"*
    android:layout_height=*"fill_parent"*

Android 应用程序设计

```
    xmlns:android="http://schemas.android.com/apk/res/android">
    <LinearLayout
        android:orientation="vertical"
        android:layout_width="fill_parent"
        android:layout_height="fill_parent">
        <TabWidget
        android:id="@android:id/tabs"
        android:layout_width="fill_parent"
        android:layout_height="wrap_content"/>
        <FrameLayout
        android:id="@android:id/tabcontent"
        android:layout_width="fill_parent"
        android:layout_height="fill_parent"
        android:padding="5dp">
            <TextView
                android:id="@+id/txt1"
                android:layout_width="fill_parent"
                android:layout_height="fill_parent"
                android:text="这是 切换卡1<tab>"/>
            <TextView
                android:id="@+id/txt2"
                android:layout_width="fill_parent"
                android:layout_height="fill_parent"
                android:text="这是 切换卡2<tab>"/>
        </FrameLayout>
    </LinearLayout>
</TabHost>
```

以上只是对 TabHost 的 ID 稍微做了修改。下面是主程序的源代码（见代码清单 2-10），读者可以对比两者之间的区别。

代码清单 2-10 第 2 章\CH2_7\src\sziit\practice\chapter2\CH2_7.java

```
public class CH2_7 extends Activity {//从活动基类 Activity 派生 CH2_7 类
    TabHost tabHost; //定义TabHost容器对象，用于存放多个Tab标签
    public void onCreate(Bundle savedInstanceState) {//派生类重载基类onCreate方法
        super.onCreate(savedInstanceState); //基类调用onCreate方法
        setContentView(R.layout.main); //设置屏幕布局，布局资源通过R.layout.main引用
        try{//防止代码运行出现异常
            tabHost=(TabHost)findViewById(R.id.tabhost_1); //获取布局文件中声明的TabHost对象
```

tabHost.setup();　//为Tabhost对象添加Tab标签
tabHost.addTab(tabHost.newTabSpec("tab1").setIndicator("TAB 1",getResources()
    .getDrawable(R.drawable.*icon*))　//设置切换卡第一个选项卡图标
    .setContent(R.id.*txt1*));　//设置切换卡第一个选项卡显示内容
tabHost.addTab(tabHost.newTabSpec("tab2").setIndicator("TAB 2",getResources()
    .getDrawable(R.drawable.*icon*))　//设置切换卡第二个选项卡图标
    .setContent(R.id.*txt2*));　//设置切换卡第二个选项卡显示内容
tabHost.setCurrentTab(0);　//设置当前的Tab标签为第一个选项卡
tabHost.setOnTabChangedListener(**new** OnTabChangeListener(){//设置切换时的事件监听
    public void onTabChanged(String tabId) {//处理Tab切换事件响应
        Toast.*makeText*(CH2_7.**this**, "切换", Toast.*LENGTH_SHORT*).show();
    }
});
}**catch**(Exception e){ //捕获异常
    e.printStackTrace();　//显示异常信息
}
}
}

2.4　项目实施

通过前面的几节，我们学习了 Android 常用的布局知识，下面就通过一个简单的项目来应用所学的知识。首先来看一下效果图，如图 2-8 所示。

图 2-8　效果图

通过图 2-8，可以分析出布局样式，充分将所学到的布局知识灵活运用。在开发过程中，还需要考虑不同分辨率的适配问题，所以在做布局的时候也要充分考虑这一点。分析效果图可以看到，自上到下分别是标题栏，然后是"安顺黄果树"、"景区地图"；第三行"实用信息"、"订单管理"，最后一行是"我的收藏"。这种效果可以通过布局间的嵌套来实现，可以通过不同的方案来实施，下面就通过相对布局的方式来实现，如代码清单 2-11 所示。读者可以通过别的方法来实现。

代码清单 2-11 就是所实现的源代码。

代码清单 2-11 布局源码

```xml
<?xml version="1.0" encoding="utf-8"?>
<RelativeLayout xmlns:android="http://schemas.android.com/apk/res/android"
    android:layout_width="fill_parent"
    android:layout_height="fill_parent"
    android:orientation="vertical"
    android:background="@drawable/bkg_img">
    <LinearLayout
        android:id="@+id/titlebar_layout"
        android:layout_width="fill_parent"
        android:layout_height="wrap_content">
        <include layout="@layout/title_bar"/>
    </LinearLayout>
    <LinearLayout
      android:layout_below="@id/titlebar_layout"
        android:layout_width="fill_parent"
        android:layout_height="wrap_content"
        android:gravity="center"
        android:layout_centerHorizontal="true"
        android:paddingTop="5dip"
        android:orientation="vertical" >
        <LinearLayout
            android:layout_width="wrap_content"
            android:layout_height="wrap_content"
           android:paddingTop="3dip"
            android:orientation="horizontal" >
            <LinearLayout
                android:layout_width="wrap_content"
                android:layout_height="wrap_content"
                android:paddingLeft="4dip"
```

```xml
            android:orientation="vertical" >
            <com.youkesf.leyou.widget.MyImageView
                android:id="@+id/localTour"
                android:layout_width="wrap_content"         android:scaleType="matrix"
                android:layout_height="wrap_content"
                android:src="@drawable/left_top" />
            <com.youkesf.leyou.widget.MyImageView
                android:id="@+id/wirelessCity"
                android:paddingTop="4dip"
                android:layout_width="wrap_content"
                android:scaleType="matrix"
                android:layout_height="wrap_content"
                android:src="@drawable/left_bottom" />
        </LinearLayout>
        <LinearLayout
            android:layout_width="wrap_content"
            android:layout_height="wrap_content"
            android:paddingLeft="4dip"
            android:orientation="vertical" >
            >
            <com.youkesf.leyou.widget.MyImageView
                android:id="@+id/mybooking"
                android:layout_width="wrap_content"         android:scaleType="matrix"
                android:layout_height="wrap_content"
                android:src="@drawable/right" />
            <com.youkesf.leyou.widget.MyImageView
                android:id="@+id/order_list"
                android:paddingTop="4dip"
                android:layout_width="wrap_content"         android:scaleType="matrix"
                android:layout_height="wrap_content"
                android:src="@drawable/right_top" />
        </LinearLayout>
    </LinearLayout>
    <RelativeLayout
        android:layout_width="wrap_content"
        android:layout_height="wrap_content"
        android:layout_margin="2dp"
```

```xml
        >
        <com.youkesf.leyou.widget.MyImageView
            android:id="@+id/favourite"
            android:layout_width="wrap_content"        android:scaleType="matrix"
            android:layout_height="wrap_content"       android:layout_margin="2dp"
            android:src="@drawable/bottom" />
        <TextView
            android:id="@+id/favourite_number"
            android:layout_alignRight="@id/favourite"
            android:layout_marginRight="10dp"
            android:layout_marginTop="5dp"
            android:textColor="@color/white"
            android:layout_width="wrap_content"
            android:layout_height="wrap_content"
            />
    </RelativeLayout>
</LinearLayout>
<RelativeLayout
    android:id="@+id/composer_buttons_wrapper"
    android:layout_width="fill_parent"
    android:layout_height="fill_parent"
    android:layout_alignParentBottom="true"
    android:layout_alignParentRight="true"
    android:clipChildren="false"
    android:clipToPadding="false"
    >
    <ImageButton
        android:id="@+id/composer_button_photo"
        android:layout_width="wrap_content"
        android:layout_height="wrap_content"
        android:layout_alignParentBottom="true"
        android:layout_alignParentRight="true"
        android:layout_marginBottom="110.667dp"
        android:layout_marginRight="10.667dp"
        android:background="@drawable/main_load"
        android:clickable="true"
        android:focusable="true"
```

```xml
            android:visibility="gone" />
        <ImageButton
            android:id="@+id/composer_button_people"
            android:layout_width="wrap_content"
            android:layout_height="wrap_content"
            android:layout_alignParentBottom="true"
            android:layout_alignParentRight="true"
            android:layout_marginBottom="97.337dp"
            android:layout_marginRight="60.667dp"
            android:background="@drawable/main_gps"
            android:clickable="true"
            android:focusable="true"
            android:visibility="gone" />
        <ImageButton
            android:id="@+id/composer_button_place"
            android:layout_width="wrap_content"
            android:layout_height="wrap_content"
            android:layout_alignParentBottom="true"
            android:layout_alignParentRight="true"
            android:layout_marginBottom="60.667dp"
            android:layout_marginRight="97.337dp"
            android:background="@drawable/main_deguide"
            android:clickable="true"
            android:focusable="true"
            android:visibility="gone" />
        <ImageButton
            android:id="@+id/composer_button_about"
            android:layout_width="wrap_content"
            android:layout_height="wrap_content"
            android:layout_alignParentBottom="true"
            android:layout_alignParentRight="true"
            android:layout_marginBottom="10.667dp"
            android:layout_marginRight="110.667dp"
            android:background="@drawable/main_about"
            android:clickable="true"
            android:focusable="true"
            android:visibility="gone" />
```

```
    </RelativeLayout>
    <RelativeLayout
        android:id="@+id/composer_buttons_show_hide_button"
        android:layout_width="60dp"
        android:layout_height="57.33333333333333dp"
        android:layout_alignParentBottom="true"
        android:layout_alignParentRight="true"
        android:visibility="gone"
        android:background="@drawable/composer_button" >
        <ImageView
            android:id="@+id/composer_buttons_show_hide_button_icon"
            android:layout_width="wrap_content"
            android:layout_height="wrap_content"
            android:layout_centerInParent="true"
            android:visibility="gone"
            android:src="@drawable/composer_icn_plus" />
    </RelativeLayout>
</RelativeLayout>
```

在项目实施中，我们采用了相对布局和线性布局相结合的方式进行。通过不断调整布局来达到美观的效果。当然，在项目的实施中也离不开美工的支持，只有 UI 设计师设计好了界面，我们才能用技术去实现布局。

2.5 技术拓展：<include>和自定义控件

在前面的源码中可能会注意到，在标题栏的布局上我们使用的是：

```
<LinearLayout
    android:id="@+id/titlebar_layout"
    android:layout_width="fill_parent"
    android:layout_height="wrap_content">
    <include layout="@layout/title_bar"/>
</LinearLayout>
```

通过<include>的方式将标题栏设置单独的通用的一个布局，提高其重用性。这也是我们在学习和工作中需要注意的一种方式。

大家也可能会注意到另外一个控件的使用：

```
<com.youkesf.leyou.widget.MyImageView
    android:id="@+id/localTour"
    android:layout_width="wrap_content"          android:scaleType="matrix"
```

android:layout_height=*"wrap_content"*
android:src=*"@drawable/left_top"* />

在这里，我们调用了一个名称为 MyImageView 的控件，通过名字来看就能够知道它并不是系统的控件，那么它是哪里来的呢？这就是自定义控件。Android 系统为我们提供的控件在日常开发中并不能很好地满足我们的需求，故需要借助已有的控件去实现一个新的控件，这就是所谓的自定义控件。

自定义控件一般是继承某个 View，重写某些方法，然后再 Layout 中引用即可。那么下面就来看一下自定义的 MyImageView 是如何定义和实现的，如代码清单 2-12 所示。

代码清单 2-12

```
public class MyImageView extends ImageView {//从基类ImageView派生出MyImageView类
    private boolean onAnimation = true;  //定义类的私有变量
    private int rotateDegree = 10;
    private boolean isFirst = true;
    private float minScale = 0.95f;
    private int vWidth;
    private int vHeight;
    private boolean isFinish = true;
    private boolean isActionMove=false;
    private boolean isScale=false;
    private Camera camera;
    boolean XbigY = false;
    float RolateX = 0;
    float RolateY = 0;
    OnViewClick onclick=null;
    public MyImageView(Context context) {//派生类构造器
        super(context);  //调用基类构造器
        camera = new Camera();  //创建Camera类对象
    }
    public MyImageView(Context context, AttributeSet attrs) {
        super(context, attrs);
        camera = new Camera();
    }
    public void SetAnimationOnOff(boolean flag) {
        onAnimation = flag;
    }
    public void setOnClickIntent(OnViewClick onclick){
        this.onclick=onclick;
```

```java
}
protected void onDraw(Canvas canvas) {//派生类重载基类重绘onDraw方法
    super.onDraw(canvas);   //基类调用onDraw方法
    if (isFirst) {
        isFirst = false;
        init();
    }
    canvas.setDrawFilter(new PaintFlagsDrawFilter(0, Paint.ANTI_ALIAS_FLAG
            | Paint.FILTER_BITMAP_FLAG));   //设置画布绘制过滤器
}
public void init() {//初始化
    vWidth = getWidth() - getPaddingLeft() - getPaddingRight();
    vHeight = getHeight() - getPaddingTop() - getPaddingBottom();
    Drawable drawable = getDrawable();
    if(drawable == null) {
        drawable = getBackground();
    }
    BitmapDrawable bd = (BitmapDrawable) drawable;
    bd.setAntiAlias(true);
    bd.setAlpha(220);
}
public boolean onTouchEvent(MotionEvent event) {//处理触摸事件
    super.onTouchEvent(event);
    if (!onAnimation)
        return true;
    switch (event.getAction() & MotionEvent.ACTION_MASK) {
    case MotionEvent.ACTION_DOWN:
        float X = event.getX();
        float Y = event.getY();
        //判断单击的坐标偏哪个方向
        RolateX = vWidth / 2 - X;
        RolateY = vHeight / 2 - Y;
        XbigY = Math.abs(RolateX) > Math.abs(RolateY) ? true : false;
        //当在这个范围的时候全部压下去
        isScale = X > vWidth / 3 && X < vWidth * 2 / 3 && Y > vHeight / 3 && Y < vHeight * 2 / 3;
        isActionMove = false;
        if (isScale) {
```

```
                    handler.sendEmptyMessage(1);
                } else {
                    rolateHandler.sendEmptyMessage(1);
                }
                break;
            case MotionEvent.ACTION_MOVE:
                float x=event.getX();
                float y=event.getY();
                if(x>vWidth || y>vHeight || x<0 || y<0){
                    isActionMove=true;
                }else{
                    isActionMove=false;
                }
                break;
            case MotionEvent.ACTION_UP:
                if (isScale) {
                    handler.sendEmptyMessage(6);
                } else {
                    rolateHandler.sendEmptyMessage(6);
                }
                break;
        }
        return true;
    }
    public interface OnViewClick {//处理单击事件
        public void onClick();
    }
    private Handler rolateHandler = new Handler() {
        private Matrix matrix = new Matrix();
        private float count = 0;
        public void handleMessage(Message msg) {//处理消息循环
            super.handleMessage(msg);
            matrix.set(getImageMatrix());
            switch (msg.what) {
                case 1:
                    count = 0;
                    BeginRolate(matrix, (XbigY ? count : 0), (XbigY ? 0 : count));
```

```
                rolateHandler.sendEmptyMessage(2);
            break;
        case 2:
            BeginRolate(matrix, (XbigY ? count : 0), (XbigY ? 0 : count));
            if (count < getDegree()) {
                rolateHandler.sendEmptyMessage(2);
            } else {
                isFinish = true;
            }
            count++;
            count++;
            break;
        case 3:
            BeginRolate(matrix, (XbigY ? count : 0), (XbigY ? 0 : count));
            if (count > 0) {
                rolateHandler.sendEmptyMessage(3);
            } else {
                isFinish = true;
                if(!isActionMove&&onclick!=null){
                    onclick.onClick();
                }
            }
            count--;
            count--;
            break;
        case 6:
            count = getDegree();
            BeginRolate(matrix, (XbigY ? count : 0), (XbigY ? 0 : count));
            rolateHandler.sendEmptyMessage(3);
            break;
        }
    }
};

private synchronized void BeginRolate(Matrix matrix, float rolateX,
        float rolateY) {//实现旋转功能
    // Bitmap bm = getImageBitmap();
```

```
            int scaleX = (int) (vWidth * 0.5f);
            int scaleY = (int) (vHeight * 0.5f);
            camera.save();
            camera.rotateX(RolateY > 0 ? rolateY : -rolateY);
            camera.rotateY(RolateX < 0 ? rolateX : -rolateX);
            camera.getMatrix(matrix);
            camera.restore();
            if (RolateX > 0 && rolateX != 0) {
                matrix.preTranslate(-vWidth, -scaleY);
                matrix.postTranslate(vWidth, scaleY);
            } else if (RolateY > 0 && rolateY != 0) {
                matrix.preTranslate(-scaleX, -vHeight);
                matrix.postTranslate(scaleX, vHeight);
            } else if (RolateX < 0 && rolateX != 0) {
                matrix.preTranslate(-0, -scaleY);
                matrix.postTranslate(0, scaleY);
            } else if (RolateY < 0 && rolateY != 0) {
                matrix.preTranslate(-scaleX, -0);
                matrix.postTranslate(scaleX, 0);
            }
            setImageMatrix(matrix);
    }

    private Handler handler = new Handler() {
        private Matrix matrix = new Matrix();
        private float s;
        int count = 0;
        public void handleMessage(Message msg) {//处理消息循环
            super.handleMessage(msg);
            matrix.set(getImageMatrix());
            switch (msg.what) {
            case 1:
                if (!isFinish) {
                    return;
                } else {
                    isFinish = false;
                    count = 0;
```

```
                    s = (float) Math.sqrt(Math.sqrt(minScale));
                    BeginScale(matrix, s);
                    handler.sendEmptyMessage(2);
                }
                break;
            case 2:
                BeginScale(matrix, s);
                if (count < 4) {
                    handler.sendEmptyMessage(2);
                } else {
                    isFinish = true;
                    if(!isActionMove && onclick!=null){
                        onclick.onClick();
                    }
                }
                count++;
                break;
            case 6:
                if (!isFinish) {
                    handler.sendEmptyMessage(6);
                } else {
                    isFinish = false;
                    count = 0;
                    s = (float) Math.sqrt(Math.sqrt(1.0f / minScale));
                    BeginScale(matrix, s);
                    handler.sendEmptyMessage(2);
                }
                break;
            }
        }
    };
    private synchronized void BeginScale(Matrix matrix, float scale) {//实现缩放
        int scaleX = (int) (vWidth * 0.5f);
        int scaleY = (int) (vHeight * 0.5f);
        matrix.postScale(scale, scale, scaleX, scaleY);
        setImageMatrix(matrix);
    }
```

```
    public int getDegree() {
        return rotateDegree;
    }
    public void setDegree(int degree) {
        rotateDegree = degree;
    }
    public float getScale() {
        return minScale;
    }
    public void setScale(float scale) {
        minScale = scale;
    }
}
```

2.6 本章小结

本章着重介绍了 Android 布局的基本知识，通过几个小的例子和一个综合性的应用来让大家了解了 Android 的 6 种布局方式，最后在技术拓展中为大家介绍了<include>和自定义控件的方法和实例。

通过本章的学习，希望大家能够使用自定义的控件，完成一个 Android 应用程序的布局。

2.7 强化练习

一、简答题

1．Android 常用布局方式有哪几种？
2．单帧布局的特点是什么？
3．切换卡的实现有两种方式，分别是_____和_____。
4．线性布局的方向有两种不同的方向属性，分别为_____和_____。
5．在表单布局中，如何设置列是隐藏还是可拉伸？

二、编程题

1．通过其他的布局方式来实现图 2-7 的布局。
2．自己实现一个自定义控件，控件能够允许用户输入内容，并单击旁边的搜索，以实现搜索功能。如图 2-9 所示。

图 2-9　自定义控件效果

第3章 Android 控件 Widgets

3.1 项目导引

前面我们介绍了 Android 的基本知识，包括布局的知识。那么如何利用我们的布局来构建我们的项目？在我们的布局文件中应填充什么？将是接下来我们需要考虑和学习的重点。

应用程序的人机交互界面由许多 Android 控件组成，在前面我们已经用到了一些常用的控件，如编辑框（EditText）、文本框（TextView）和按钮（Button）等。这些控件是直接与用户交互的对象，掌握好这些控件的使用对我们在以后的开发中会起到至关重要的作用。本章我们将详细介绍一下这些基本的控件，并结合具体的实例来加深对它们的了解。

前面我们已经掌握了 Android 应用程序的基本组件和它的生命周期，下面我们就开始逐步进行 Android 应用程序的开发。首先，先来学习关于用户界面开发的知识，这对于使用应用程序创建用户界面是至关重要的。一个良好的用户界面的必备条件是内容清楚、指示明白、屏幕美观和具有亲切感。应用程序界面的设计就是对控件进行恰当地取舍以及功能的选择和处理的过程。

本章我们将学习基本的 Android UI 元素，并了解如何使用各个控件来创建我们想要的用户界面。我们会详解 Android 中的基本的用户控件，读者可以把此章作为以后在用户界面设计上的参考手册，在开发中可以查询基本控件的使用方法。

3.2 项目分析

Android 用户界面的开发主要包括两个方面的知识：用户界面的设计和事件处理。在 Android 中，用户界面是由 View 和 ViewGroup 对象构建的，View 与 ViewGroup 都有很多种类，而它们都是 View 类的子类；事件处理则包括按钮事件、触屏事件以及一些高级控件的事件监听。在这一节中我们重点学习一下用户界面的设计，事件处理的相关知识我们在后面中会讲解。

第 3 章 Android 控件 Widgets

1. 用户界面的生成

Android 用户界面有两种生成方式，一种是通过 xml 布局文件来生成，另一种是用代码直接生成。对于 xml 生成的布局文件我们可以通过 ADT 提供的 UI 预览功能来预览所创建的用户界面。在写好布局的 xml 文件后只要打开项目中的"/res/layout/main.xml"并右键单击，依次选择"Open With" → "Android Layout Editor"菜单命令即可；也可以直接双击打开布局文件，然后切换到 UI 设计界面，如图 3-1 所示。

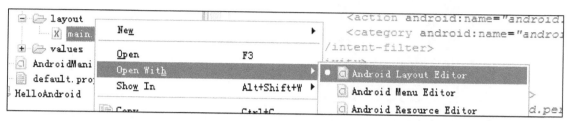

图 3-1 使用 Android Layout Editor 打开布局文件

打开后如图 3-2 所示，左边的 Layouts 标签的内容是一些线性布局，可以使用它来完成对布局的排版，如横向或者纵向。Views 标签则是 UI 控件，可以将这些控件直接拖到右边的窗口中进行编辑。

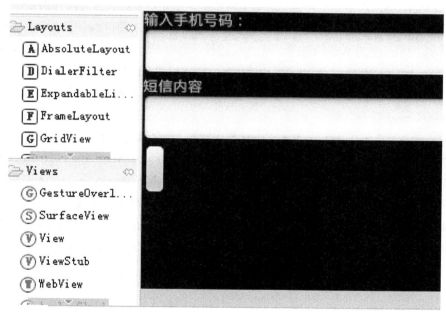

图 3-2 打开布局文件

我们可以通过 xml 编辑器和上面的功能配合使用来完成对布局的设计和开发。在这里向大家推荐一款开源的软件 DroidDraw，它是一款公开了源码的 UI 设计器，用户可以根据自己的需要来进行修改。该软件的功能比较强大，可以直接拖动控件到窗口，然后设置属性、参数等。设置好参数以后可以单击"Generate"按钮来生成相应的布局文件。当然我们也可以单击"Load"按钮来载入已经编辑好的布局文件。用户界面如图 3-3 所示。

63

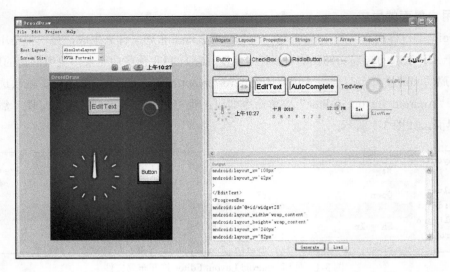

图 3-3 DroidDraw 用户界面

2. 用户界面的设计

（1）View 简介

View 对象是 Android 平台中体现用户界面的基础单位。任何一个 View 对象都将继承 android.view.View 类，它是一个存储了屏幕上特定的一个矩形布局和内容属性的数据结构。一个 View 对象可以处理测距、布局、绘图、焦点变换、滚动条以及屏幕区域自己表现的按键和手势。作为一个基类，View 类为 Widget 服务，Widget 则是一组用于绘制交互屏幕元素的完全实现子类。Widget 处理自己的测距和绘图，所以可以快速地用它们去构建 UI。常用到的 Widget 包括 Text View、EditText、Button、单选按钮（RadioButton）、复选框（Checkbox）和图片按钮（Image Button）等。

（2）ViewGroup 简介

ViewGroup 是一个 android.view.ViewGroup 类的对象。顾名思义，ViewGroup 是一个特殊的 View 对象，它的功能是装载和管理一组下层的 View 和其他 ViewGroup。ViewGroup 可以为 UI 增加结构，并且将复杂的屏幕元素组成一个独立的实体。作为一个基类，ViewGroup 为 Layout 服务，Layout 则是一组提供屏幕界面通用类型的完全实现子类。Layout 可以为一组 View 构建一个结构。

图 3-4 所示是一个由 View 和 ViewGroup 布局的活动（Activity）界面。从图中我们可以看到一个 Activity 的界面可以包含多个 ViewGroup 和 View，通过两者的组合使用能够更好地完成更复杂的界面的设计。

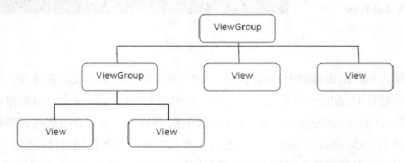

图 3-4 View 与 ViewGroup 组合使用布局的 Activity 界面

第 3 章 Android 控件 Widgets

一个新的 Activity 被创建的时候是一个空白屏幕，我们可以把自己的用户界面放到上面。要设置用户界面，可以调用 setContentView()，并传入需要显示的 View 实例（通常是一个布局）。由于空白屏幕并不是我们所想要的，所以在创建一个新的 Activity 时，在 onCreate()处理程序中总是采用调用 setContentView()的方法来设置我们需要显示的用户界面。

在本节的开始我们就提到了可以有两种方法来给我们的 Activity 设置用户界面。代码清单 3-1 演示了用布局文件设置 Activity 界面的方法。

代码清单 3-1 用布局文件设置 Activity 的界面

```
public void onCreate(Bundle savedInstanceState) {//重写基类onCreate方法
    super.onCreate(savedInstanceState); //调用基类onCreate方法
    setContentView(R.layout.main); //显示用户界面布局，布局资源通过 R.layout.main 引用
    TextView tv=(TextView)findViewById(R.id.TextView1); //应用布局文件中所定义的 View
}
```

在代码清单 3-1 中，通过 setContentView(R.layout.main)来将布局文件 main.xml 设置成为应用程序的布局，并通过 findViewById(R.id.TextView1)来引用布局文件中所使用的 View 控件。注意：通过布局文件来设置 Activity 的界面可以将表示层和应用程序层逻辑分开，这样就提供了无需修改代码就可以修改表示层的灵活方法。这样也为不同的硬件设置指定不同的优化布局提供可能，并可以根据硬件状态的变化在运行时修改这些布局。

代码清单 3-2 演示了用代码的方式来给 Activity 设置界面。在代码清单 3-2 中为 Activity 创建了一个新的 TextView 作为用户界面。

代码清单 3-2 用代码设置 Activity 界面

```
public void onCreate(Bundle savedInstanceState) {//重写基类onCreate方法
    super.onCreate(savedInstanceState); //调用基类onCreate方法
    TextView tv= new TextView(); //创建新的TextView控件
    setContentView(tv); // 将新的TextView用作用户界面
    tv.setText("HELLO!"); //在界面中显示"HELLO!"信息
}
```

setContentView()方法接收的是一个单独的 View 实例，所以我们需要把多个控件归类成一组从而保证可以使用一个 View 或者 ViewGroup 来引用一个布局。

3.3 技术准备

3.3.1 知识点 1：文本框(TextView)

在前面的学习中我们已经知道 TextView 是用来显示文本信息的控件。我们可以在布局文件中设置 TextView 的内容，当然也可以在程序中动态地设置 TextView 的内容。下面这个例子中我们通过定义格式化的定型对象来对 TextView 中的内容进行样式化，这相当于 CSS 样式的方法，可以用来指定颜色、大小等。源码详见本书源代码：第 3 章\CH3_1。程序运行如图 3-5 所示。

图 3-5　TextView 使用

在程序中我们使用了两个 TextView，对第一个 TextView 我们通过 style 来设置 TextView 文本的字体大小和字体颜色信息。文本内容由布局文件 main.xml 通过"android:text="来设置。对第二个 TextView 的文字内容和字体颜色以及背景颜色在程序中来设置。下面我们来看一下布局文件 main.xml 文件的代码清单。

代码清单 3-3　第 3 章\CH3_1\res\layout\main.xml

```xml
<?xml version="1.0" encoding="utf-8"?>
<LinearLayout xmlns:android="http://schemas.android.com/apk/res/android"
    android:orientation="vertical"
    android:layout_width="fill_parent"
    android:layout_height="fill_parent"
    >
<TextView
    style="@style/style1"
    android:layout_width="fill_parent"
    android:layout_height="wrap_content"
    android:text="@string/teststyle"
    />
<TextView
    android:id="@+id/textview2"
```

```
        android:layout_width="wrap_content"
        android:layout_height="wrap_content"
        />
</LinearLayout>
```

在上面的代码中我们需要注意:在第一个 TextView 中 style= "@style/style1"这一行我们通过 style 来设置 TextView 的属性。其中,style.xml 是事先定义好的样式,位于 res\values 下。Style.xml 的源代码如代码清单 3-4 所示。

代码清单 3-4 第 3 章\CH3_1\res\values\style.xml

```
<?xml version="1.0" encoding="UTF-8"?>
<resources>
    <style name="style1">
        <item name="android:textSize">40sp</item>
        <item name="android:textColor">#ECA248</item>
    </style>
</resources>
```

在看完布局文件和样式文件以后我们来分析一下主程序的代码,如代码清单 3-5 所示。

代码清单 3-5 第 3 章\CH3_1\src\sziit\practice\chapter3\CH3_1

```
//包的引入语句省略
public class CH3_1 extends Activity {//从基类Activity派生子类CH3_1
    public void onCreate(Bundle savedInstanceState) {//重写基类onCreate方法
        super.onCreate(savedInstanceState);  //调用基类on Create方法
        setContentView(R.layout.main);  //设置屏幕布局,布局资源通过R.layout.main引用
        TextView tv2=(TextView)findViewById(R.id.textview2);  //引用布局中的TextView对象
        tv2.setBackgroundColor(Color.RED);  //设置TextView对象的背景色
        tv2.setTextSize(20);  //设置TextView对象显示的字体大小
        tv2.setTextColor(Color.BLUE);  //设置TextView对象显示的文本颜色
        tv2.setText("在程序中设置的文字内容并设置文字颜色,大小和背景颜色!");
    }//设置TextView对象显示的内容信息
}
```

在上面的代码中需要注意:首先要通过 findViewById(R.id.textview2)获得布局文件中 id 为 textview2 的 TextView,然后通过 setBackgroundColor(Color.RED)来将 TextView 的背景颜色设置为红色,通过 setTextSize(20)将文字的字号设置为 20,通过 setTextColor(Color.BLUE)将文字颜色设置为蓝色,最后通过 setText()来设置 TextView 的内容。在上面的例子中,我们使用了两种不同的方式设置了 TextView 的属性,在实际的开发过程中可以根据不同的情况来使用不同的方法。

在上面的代码中我们使用了 Color.BLUE 来将字体颜色设置成为蓝色,在 Android 中有 12 种不同的常用颜色:

Color.BLACK——黑色

Color.BLUE——蓝色

Color.CYAN——青绿色

Color.DKGRAY——深灰色

Color.GRAY——灰色

Color.GREEN——绿色

Color.LTGRAY——浅灰色

Color.MAGENTA——紫红色

Color.RED——红色

Color.TRANSPARENT——透明的

Color.WHITE——白色

Color.YELLOW——黄色

这些颜色常数通常定义在 Android.graphics.Color 中，如表 3-1 所示。

表 3-1　　　　　　　　　　　　　　Android 颜色常数

类型	常数	值	色码
int	BLACK	−16777216	0xff000000
int	BLUE	−16776961	0xff0000ff
int	CYAN	−16711681	0xff00ffff
int	DKGRAY	−12303292	0xff444444
int	GRAY	−7829368	0xff888888
int	GREEN	−16711936	0xff00ff00
int	LTGRAY	−3355444	0xffcccccc
int	MAGENTA	−65281	0xffff00ff
int	RED	−65536	0xffff0000
int	TRANSPARENT	0	0x00000000
int	WHITE	−1	0xffffffff
int	YELLOW	−256	0xffffff00

除了表 3-1 中给出的颜色常数，在下面的表 3-2 中我们还列出了几个常用的 Text View 对象方法与其对应的 xml 属性。

表 3-2　　　　　　　　　　TextView 对象方法与对应的 xml 属性

TextView 对象方法	xml 属性
setTextColor	android:textColor
setTextSize	android:textSize
setText	android:text
setBackgroundResource	android:background
setHeight/setWidth	android:height/android:width

3.3.2 知识点 2：编辑框（EditText）

EditText 也是我们经常用到的控件。顾名思义，EditText 就是能够编辑 Text 的文本框，我们可以在里面输入内容，它可以接受多行输入并能自动换行。下面同样通过一个实例来向大家演示一下 EditText 的基本用法。实例源码详见本书所附源代码：第 3 章\CH3_02。我们先来看一下程序的运行结果，如图 3-6 和图 3-7 所示，在用户没有输入用户名和用户密码时则会提示用户输入用户名和密码，当用户输入用户名和密码时会在下面的 TextView 中显示出用户的输入。

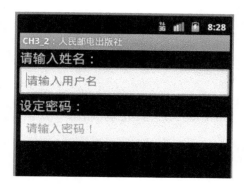
图 3-6　编辑框（EditText）实例

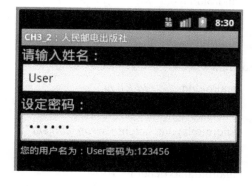
图 3-7　获取编辑框（EditText）中的内容

在这个例子中我们在布局文件 main.xml 中定义了两个 EditText 来作为用户输入的文本框，3 个 TextView 用来输出提示信息。其中，在第 2 个 EditText 控件上我们设置了事件监听方法：**setOnKeyListener**()，并实现了 **onKey**()方法，当用户在密码框中按键输入时便会触发这个事件，从而可以通过 getText()方法来取得用户输入的内容。

下面我们还是先来看一下布局文件的代码清单，如代码清单 3-6 所示。

代码清单 3-6　第 3 章\CH3_2\res\layout\main.xml

```
<?xml version="1.0" encoding="utf-8"?>
<LinearLayout xmlns:android="http://schemas.android.com/apk/res/android"
    android:orientation="vertical"
    android:layout_width="fill_parent"
    android:layout_height="fill_parent"
    >
<TextView
    android:layout_width="fill_parent"
    android:layout_height="wrap_content"
    android:text="@string/name"
    />
<EditText
```

```
    android:id="@+id/edtName"
    android:layout_width="fill_parent"
    android:layout_height="wrap_content"
    android:maxLength="10"
    android:hint="请输入用户名"
    android:textColorHint="@drawable/red"
    />
<TextView
    android:layout_width="fill_parent"
    android:layout_height="wrap_content"
    android:text="@string/password"/>
<EditText
    android:id="@+id/PWD"
    android:layout_width="fill_parent"
    android:layout_height="wrap_content"
    android:password="true"
    android:maxLength="10"/>
<TextView
    android:id="@+id/message"
    android:layout_width="wrap_content"
    android:layout_height="wrap_content"
    />
</LinearLayout>
```

在上面的代码中我们需要注意一下如下代码：

android:maxLength="10"

android:hint="请输入用户名"

android:textColorHint="@drawable/red"

我们通过 android:maxLength="10" 来设定用户名的长度最大为 10 个字符；android:hint="请输入用户名" 的作用是当用户名的编辑框中没有输入内容时就会显示提示信息，在密码的编辑框中我们在程序中通过 setHint() 来实现同样的效果；android:textColorHint="@drawable/red" 用来设置提示显示信息的颜色。这里的颜色文件位于第 3 章\CH3_02\res\values 下的 color.xml 文件。color 文件的源代码如代码清单 3-7 所示，在代码中我们只设定了一个颜色标签。

代码清单 3-7 第 3 章\CH3_2\res\values\color.xml

```
<?xml version="1.0" encoding="UTF-8"?>
<resources>
    <drawable name="red">#FFFF0000</drawable>
</resources>
```

布局文件完成后下面我们来看一下具体的代码实现。代码如代码清单3-8所示。

代码清单3-8 第3章\CH3_2\src\sziit\practice\chapter3\CH3_2.java

```
public class CH3_2 extends Activity {//从基类Activity派生子类CH3_2
    //声明TextView、EditText对象
    private TextView tv；//定义TextView对象
    private EditText edtName；//定义EditText对象，用于输入用户名
    private EditText edtPWD；//定义EditText对象，用于输入密码
    public void onCreate(Bundle savedInstanceState) {//子类重写基类onCreate方法
        super.onCreate(savedInstanceState);　//调用基类onCreate方法
        setContentView(R.layout.main)；//设置屏幕布局，通知R.layout.main引用布局资源
        //获得TextView和EditView对象
        tv=(TextView)findViewById(R.id.message)；//通过id.message引用布局中的TextView对象
        edtName=(EditText)findViewById(R.id.edtName)；//通过id.edtName引用TextView对象
        edtPWD=(EditText)findViewById(R.id.PWD)；//通过R.id.PWD引用TextView对象
        //设置提示信息，当密码为空时提示
        //用户名为空的提示是在xml中通过：android:hint="请输入用户名"实现
        edtPWD.setHint("请输入密码！");
        edtPWD.setOnKeyListener(new EditText.OnKeyListener(){//设置事件监听方法
            public boolean onKey(View v, int keyCode, KeyEvent event) {//实现onKey
                tv.setText("您的用户名为："+edtName.getText().toString()+"密码为:"+edtPWD.getText().toString());
                return false;
            }
        });
    }
}
```

在上面的代码中我们使用 **setOnKeyListener**()来监听用户在编辑框中的输入。我们重写了 **onKey**()方法，当用户输入内容时我们将EditText中的文本信息通过**getText**()方法获得，并显示在TextView中。

这个例子中的实时输入实时显示的效果可以扩展到许多的手机应用中，如文字过滤效果：例如，当用户输入不雅文字时可以不接受部分关键字，如用户输入"shit"时在TextView中显示sh*t。有兴趣的读者可以自己尝试一下。

此外，不仅仅是Widget才具备setOnKeyListener方法的重写功能，事实上，在View里也有View.setOnKeyListener()，也就是捕捉用户单击键盘时的事件处理，但需要注意：只有View取得用户焦点时才能触发onKeyDown事件。

3.3.3 知识点3：按钮（Button）和图片按钮（ImageButton）

按钮是我们在开发中最常用到的控件，在 Android 平台中，按钮是通过 Button 来实现的。Button 的属性与前面我们讲解过的 TextView 和 EditView 相似，大家可以参照前面的知识来掌握 Button 的属性。

为了能响应按钮被按下后的事件，需要对按钮设置 setOnClickListener()事件监听。本节的实例我们将通过按钮来改变 TextView 中文字的颜色。源代码见第 3 章\CH3_3。程序运行后效果如图 3-8 和图 3-9 所示，图 3-8 所示是程序初始的状态，图 3-9 所示是单击按钮后的效果。

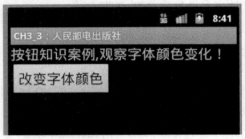
图 3-8　程序初始状态

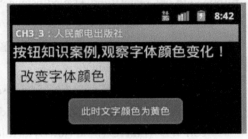
图 3-9　单击"开始"按钮

程序的布局很简单，我们这里不再介绍布局文件的源代码，直接来看程序的代码实现，如代码清单 3-9 所示。

代码清单 3-9　第 3 章\CH3_3\src\sziit\practic\chapter3\CH3_3.java

```
public class CH3_3 extends Activity {//从基类Activity派生子类CH3_3
    //声明TextView和Button对象
    private Button btnStart; //定义私有Button对象
    private TextView tv; //定义私有TextView对象
    private int[] colorArray; //定义私有int[]数组
    private int colorNum; //定义私有int变量
    private String strColor[]; //定义私有String对象
    public void onCreate(Bundle savedInstanceState) {//子类重写基类的onCreate方法
        super.onCreate(savedInstanceState); //调用基类的onCreate方法
        setContentView(R.layout.main); //设置用户布局，通过R.layout.main引用布局资源
        btnStart=(Button)findViewById(R.id.btnStart); //通过findViewById()获得Button对象
        tv=(TextView)findViewById(R.id.texview1); //通过findViewById()获得TextView对象
        //设置颜色数组
        colorArray=new int[]{Color.YELLOW,Color.BLUE,Color.GREEN,Color.DKGRAY,Color.CYAN,Color.MAGENTA};
        strColor=new String[]{"黄色","蓝色","绿色"," 深灰色"," 青绿色"," 紫红色"};
        colorNum=0;
        btnStart.setOnClickListener(new Button.OnClickListener(){//设置按钮的监听方法
            public void onClick(View v){ //处理单击（Click）事件响应
```

```
        if(colorNum<colorArray.length){
            tv.setTextColor(colorArray[colorNum]);
            Toast toast=Toast.makeText(CH3_3.this, "此时文字颜色为"+strColor[colorNum], Toast.LENGTH_SHORT);
            toast.setGravity(Gravity.TOP, 0 150);  //设置Toast显示的位置
            toast.show();  //显示提示信息
            colorNum++;
        }
        else colorNum=0;
    }
});
    }
}
```

在程序中我们定义了一个 TextView 对象和 Button 对象。两个颜色数组 colorArray 和 strColor, 前者用来在程序中设定 TextView 中文字的颜色，后者在 Toast 的提示信息中使用，并通过使用 colorNum 来控制数组的下标从而更新字体颜色。

为了能更新 TextView 中的字体颜色我们设置 setOnClickListener()来监听每个按钮单击（Click）事件，通过 setTextColor()方法将颜色数组 colorArray 中的颜色设置成 TextView 中字体的颜色；在 Toast 的提示信息中我们则是将取到的字符串数组中的颜色信息来呈现给用户。其中注意 toast.setGravity(Gravity.TOP, 0, 150)，我们用它来设定提示信息的显示位置，默认的情况下是在屏幕的下端，现在让它显示在屏幕的上端，其中最后两个参数分别是指在 x 轴和 y 轴上的位移。关于 Toast 的详细知识我们将在下节讲解。

图片按钮（ImageButton），顾名思义是带有图片的按钮，其实现的功能与按钮实现的功能大体相同，带有图片的按钮能够让我们的程序更加美观。图片按钮显示一个可以被用户单击的图片外观按钮，默认情况下，ImageButton 看起来像一个普通的按钮，在不同状态（如按下）下改变背景颜色。按钮的图片可用通过 xml 元素的 android:src 属性或 setImageResource(int)方法指定。如果要删除按钮的背景，可以自定义背景图片或设置背景为透明。同时，Android 系统为我们提供了默认的按钮选中时的状态。

为了表示不同的按钮状态（焦点、选择等），可以为各种状态定义不同的图片。例如，定义蓝色图片为默认图片，黄色图片为获取焦点时显示的图片，黄色图片为按钮被按下时显示的图片。一个简单的方法可以做到这点——通过 xml 的 "selector." 配置，如下：

```xml
<?xml version="1.0" encoding="utf-8"?>
<selector xmlns:android="http://schemas.android.com/apk/res/android">
    <item android:state_pressed="true"
          android:drawable="@drawable/button_pressed" /> <!-- pressed -->
    <item android:state_focused="true"
          android:drawable="@drawable/button_focused" /> <!-- focused -->
    <item android:drawable="@drawable/button_normal" /> <!-- default -->
</selector>
```

保存上面的 xml 到 res/drawable/文件夹下，文件名我们暂且命名为 myselector.xml，然后将该文件名作为一个参数设置到 ImageButton 的 android:src 属性为 "@drawable/myselector"。Android 根据按钮的状态改变会自动地去 xml 中查找相应的图片以显示。

元素的顺序很重要，因为根据这个顺序可以判断是否适用于当前按钮状态，这也解释了为什么正常（默认）状态指定的图片放在最后，因为它只会在 pressed 和 focused 都判断失败之后才会被采用。例如，按钮被按下时是同时获得焦点的，但是获得焦点并不一定按了按钮，所以这里会按顺序查找，找到合适的就不往下找了。这里按钮被单击了，那么在 myselector.xml 文件中定义的第一个图片将被选中，且不再在后面查找其他状态。读者可以自己实现此实例。

3.3.4　知识点 4：复选框（CheckBox）和单选按钮（RadioButton）

单项选择和多项选择大家都应该不陌生。我们在日常的学习和生活中都接触过，在 Android 中同样提供了单项选择和多项选择组件。单项选择，顾名思义就是只能选择其中一项，在通常的应用中是单选按钮（RadioButton）和 RadioGroup 一起使用。在单选按钮没有被选中时，用户能够通过按下或单击来选中它，用户一旦选中就不能够取消。当一个单选组（RadioGroup）包含几个单选按钮时，选中其中一个的同时将取消其他选中的单选按钮。而多选按钮（CheckBox）则与 RadioButton 不同，它可以选中也可以不选中，也可以选择多个选型。在多项选择中我们需要注意，既然用户可以选择多个选项，那么就要对每个选项进行事件监听来确定用户是否选择了某一项。

为了能够更直观地了解 RadioButton、RadioGroup 和 CheckBox 的应用，我们通过实例 3_4（代码详见第 3 章\CH3_4）来演示它们的用法。在这个实例中我们模拟了一个用户注册的模块，用户需要填写用户名，然后选择性别、选择自己的爱好，样式如图 3-10 所示。

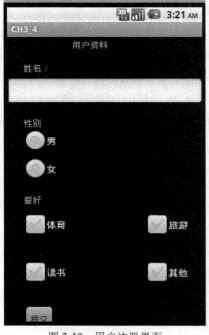

图 3-10　用户注册界面

注册信息中的每项都是必填的，如果没有填写的信息，则会通过上节学的 Toast 来向用户提示。相应地，当用户选择选项时我们也会用 Toast 来向用户提示。用户填写完信息后单击"提交"按钮，将用户填写和选择的信息通过 Intent 传递给下一个 Activity，并在 TextView 中显示用户信息，如图 3-11 和图 3-12 所示。

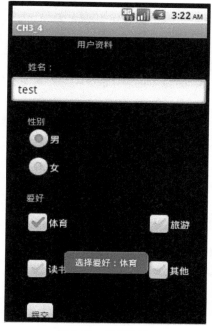

图 3-11　用户注册　　　　　　　　　　图 3-12　注册完成

下面首先来看一下布局文件，如代码清单 3-10 所示。

代码清单 3-10　第 3 章\ CH3_4\res\layout\main.xml

```
<?xml version="1.0" encoding="utf-8"?>
<AbsoluteLayout android:id="@+id/widget0"
    android:layout_width="fill_parent" android:layout_height="fill_parent"
    xmlns:android="http://schemas.android.com/apk/res/android">
    <TextView
        android:id="@+id/tvInfromation"
        android:layout_width="110px"
        android:layout_height="26px"
        android:text="用户资料"
        android:layout_x="80px"
        android:layout_y="2px">
    </TextView>
    <TextView
        android:id="@+id/ tvName "
```

```xml
            android:layout_width="wrap_content"
            android:layout_height="20px"
            android:text="姓名："
            android:layout_x="20px"
            android:layout_y="32px">
</TextView>
<EditText
            android:id="@+id/userName"
            android:layout_width="fill_parent"
            android:layout_height="wrap_content"
            android:layout_x="0px"
            android:layout_y="52px">
</EditText>
<TextView android:id="@+id/sex"
            android:layout_width="wrap_content"
            android:layout_height="wrap_content"
            android:text="性别"
            android:layout_x="20px"
            android:layout_y="102px">
</TextView>
<RadioGroup
            android:id="@+id/radioGroup1"
            android:layout_width="wrap_content"
            android:layout_height="wrap_content"
            android:layout_x="20px"
            android:layout_y="115px">
    <RadioButton
            android:id="@+id/male"
            android:layout_width="wrap_content"
            android:layout_height="wrap_content"
            android:text="男" >
    </RadioButton>
    <RadioButton
            android:id="@+id/female"
            android:layout_width="wrap_content"
            android:layout_height="wrap_content"
            android:text="女" >
```

```xml
        </RadioButton>
    </RadioGroup>
    <TextView
        android:layout_width="wrap_content"
        android:layout_height="wrap_content"
        android:text="爱好"
        android:layout_x="20px"
        android:layout_y="200px">
    </TextView>
    <CheckBox
        android:id="@+id/sports"
        android:layout_width="wrap_content"
        android:layout_height="wrap_content"
        android:text="体育"
        android:layout_x="20px"
        android:layout_y="222px">
    </CheckBox>
    <CheckBox
        android:id="@+id/travelling"
        android:layout_width="wrap_content"
        android:layout_height="wrap_content"
        android:text="旅游"
        android:layout_x="170px"
        android:layout_y="222px">
    </CheckBox>
    <CheckBox
        android:id="@+id/reading"
        android:layout_width="wrap_content"
        android:layout_height="wrap_content"
        android:text="读书"
        android:layout_x="20px"
        android:layout_y="282px">
    </CheckBox>
    <CheckBox
        android:id="@+id/others"
        android:layout_width="wrap_content"
        android:layout_height="wrap_content"
```

```
                android:text="其他"
                android:layout_x="170px"
                android:layout_y="282px">
        </CheckBox>
        <Button android:id="@+id/submit"
                android:layout_width="wrap_content"
                android:layout_height="wrap_content"
                android:text="提交"
                android:layout_x="20px"
                android:layout_y="342px">
        </Button>
</AbsoluteLayout>
```

在布局文件中，我们使用了一个 RadioGroup，里面包含两个 RadioButton 来构成性别的选项。在 RadioGroup 中的 RadioButton 中只能有一个是可以选择的，即当一个选择时另一个不能同时被选中。我们可以用单项选择题的形式来理解，RadioGroup 里面的 RadioButton 构成了选择题的选项，而一个单选题中的答案中只能选择一个。在布局文件中我们同样设置了 4 个 CheckBox 按钮来供用户选择自己的爱好。与 RadioButton 不同，CheckBox 可以选择多个选项。另外还需要注意的是，android:layout_x="20px",android:layout_y="342px"分别表示距离原点的坐标位置。

在完成了对界面的布局之后，我们来看一下主程序中的实现，程序代码如代码清单 3-11 所示。

代码清单 3-11 第 3 章\ CH3_4\ sziit\practice stady\chapter3\CH3_4.java

```
//包引入语句省略
public class CH3_4 extends Activity {//从Activity基类派生子类CH3_4
    private RadioGroup radioGroup1；//定义私有RadioGroup对象
    private RadioButton rbSex1；//定义私有单选按钮RadioButton对象
    private RadioButton rbSex2；// 定义私有单选按钮RadioButton对象
    // 定义多选按钮CheckBox对象
    private CheckBox cbSport；//定义私有CheckBox对象
    private CheckBox cbTravelling；//定义私有CheckBox对象
    private CheckBox cbReading；//定义私有CheckBox对象
    private CheckBox cbOthers；//定义私有CheckBox对象
    private EditText etName；// 定义私有编辑框对象
    private Button btnOK；// 定义私有按钮对象
//定义字符串变量，userName、sex和hobby分别用来表示姓名、性别和爱好。
    String userName;
    String sex="";
```

```java
String hobby="";
public void onCreate(Bundle savedInstanceState) {//子类重写基类的onCreate方法
    super.onCreate(savedInstanceState);  //调用基类的onCreate方法
    setContentView(R.layout.main);  //设置布局文件，通过R.layout.main引用布局资源
    //从布局文件中获取各个对象
    radioGroup1=(RadioGroup)findViewById(R.id.radioGroup1);  //引用布局中的RadioGroup
    rbSex1=(RadioButton)findViewById(R.id.male);  //通过findViewById引用RadioButton
    rbSex2=(RadioButton)findViewById(R.id.female);  //通过findViewById引用RadioButton
    cbSport=(CheckBox)findViewById(R.id.sports);  //通过findViewById引用CheckBox
    cbTravelling=(CheckBox)findViewById(R.id.travelling);  //通过findViewById引用CheckBox
    cbReading=(CheckBox)findViewById(R.id.reading);  //通过findViewById引用CheckBox
    cbOthers=(CheckBox)findViewById(R.id.others);  //通过findViewById引用CheckBox
    btnOK=(Button)findViewById(R.id.submit);  //通过findViewById引用Button
    etName=(EditText)findViewById(R.id.userName);  //通过findViewById引用EditText
    //设置radioGroup1的事件监听
    radioGroup1.setOnCheckedChangeListener(new RadioGroup.OnCheckedChangeListener(){
        public void onCheckedChanged(RadioGroup group, int checkedId) {//处理单击变更
            if(checkedId==rbSex1.getId()){//通过Id识别RadioButton对象rbSex1
                Toast.makeText(CH3_4.this, "性别："+rbSex1.getText(),
                    Toast.LENGTH_SHORT).show();  //显示性别信息
            } else if(checkedId==rbSex2.getId()){//通过Id识别RadioButton对象rbSex2
                Toast.makeText(CH3_4.this, "性别："+rbSex2.getText(),
                    Toast.LENGTH_SHORT).show();  //显示性别信息
            }
        }
    });
    //设置CheckBox的事件监听,因为我们需要记录每个复选框的选择情况,
    //所以我们这里需要对4个CheckBox都要设置监听
    cbSport.setOnCheckedChangeListener(new CheckBox.OnCheckedChangeListener(){
        public void onCheckedChanged(CompoundButton arg0, boolean arg1) {
            if(cbSport.isChecked()){//检查体育爱好CheckBox对象cbSport状态
                hobby+=cbSport.getText().toString()+"、";
                Toast.makeText(CH3_4.this, "选择爱好："+cbSport.getText(),
                    Toast.LENGTH_SHORT).show();  //显示提示信息
            }else {
                hobby+="";
                Toast.makeText(CH3_4.this, "取消选择
```

Android 应用程序设计

```
"+cbSport.getText(),Toast.LENGTH_SHORT).show();  //显示提示信息
                }
            }
        });
        cbTravelling.setOnCheckedChangeListener(new CheckBox.OnCheckedChangeListener(){
            public void onCheckedChanged(CompoundButton buttonView,
                boolean isChecked) {//处理单击变化事件响应
                if(cbTravelling.isChecked()){//检查旅游爱好CheckBox对象cbTravelling状态
                    hobby+=cbTravelling.getText().toString()+"、";
                    Toast.makeText(CH3_4.this, "选择爱好："+cbTravelling.getText(),
Toast.LENGTH_SHORT).show();  //用Toast显示提示信息
                } else{
                    hobby+="";
                    Toast.makeText(CH3_4.this, "取消选择"+cbTravelling.getText(),
Toast.LENGTH_SHORT).show();  //用Toast显示提示信息
                }
            }
        });
        cbReading.setOnCheckedChangeListener(new CheckBox.OnCheckedChangeListener(){
            public void onCheckedChanged(CompoundButton buttonView,
                boolean isChecked) {//处理单击变化事件响应
                if(cbReading.isChecked()){//检查阅读爱好CheckBox对象cbReading状态
                    hobby+=cbReading.getText().toString()+"、";
                    Toast.makeText(CH3_4.this, "选择爱好："+cbReading.getText(),
Toast.LENGTH_SHORT).show();  //用Toast显示提示信息
                } else{
                    Toast.makeText(CH3_4.this, "取消选择"+cbReading.getText(),
Toast.LENGTH_SHORT).show();  //用Toast显示提示信息
                }
            }
        });
        cbOthers.setOnCheckedChangeListener(new CheckBox.OnCheckedChangeListener(){
            public void onCheckedChanged(CompoundButton buttonView,
                boolean isChecked) {//处理单击变化事件响应
                if(cbOthers.isChecked()){//检查其他爱好CheckBox对象cbOthers状态
                    hobby+=cbOthers.getText().toString()+"、";
                    Toast.makeText(CH3_4.this, "选择"+cbOthers.getText(),
```

```java
Toast.LENGTH_SHORT).show();  //用Toast显示提示信息
                } else {
                    hobby+="";
                    Toast.makeText(CH3_4.this, "取消选择"+cbOthers.getText(),Toast.LENGTH_SHORT).show();  //用Toast显示提示信息
                }
            }
        });
        //设置提交按钮事件监听
        btnOK.setOnClickListener(new Button.OnClickListener(){
            public void onClick(View arg0) {//处理按钮(Button)单击事件响应
                if(etName.getText().length()==0){
                    Toast.makeText(CH3_4.this, "请输入姓名！",Toast.LENGTH_SHORT).show();  //用Toast显示提示信息
                }else if(radioGroup1.getCheckedRadioButtonId()!=rbSex1.getId()&&radioGroup1.getCheckedRadioButtonId()!=rbSex2.getId()){
                    Toast.makeText(CH3_4.this, "请选择性别!",Toast.LENGTH_SHORT).show();  //用Toast显示提示信息
                }else if(!cbSport.isChecked()&&!cbTravelling.isChecked()&&!cbReading.isChecked()&&!cbOthers.isChecked()){
                    Toast.makeText(CH3_4.this, "请至少选择一个爱好!",Toast.LENGTH_SHORT).show();  //用Toast显示提示信息
                }else {
                    if(rbSex1.isChecked()){
                        sex="男性";
                    }else {
                        sex="女性";
                    }
                    userName=etName.getText().toString();
                    //创建一个新的意图(Intent)对象，并指定其class
                    Intent intent=new Intent();
                    intent.setClass(CH3_4.this, CH3_4_01.class);
                    Bundle bundle=new Bundle();  //新建Bundle对象，并将要传入的数据传入
                    bundle.putString("name", userName);
                    bundle.putString("sex", sex);
```

```
                    bundle.putString("hobby", hobby);
                    intent.putExtras(bundle);     //将bundle对象赋给Intent
                    startActivity(intent);  //调用第二个Activity，CH3_4_01;
                }
            }
        });
    }
}
```

在主程序中我们对 RadioGroup 和 CheckBox 设置事件监听 setOnCheckedChangeListener()，来监听按钮的选择情况。在 RadioGroup 的 setOnCheckedChangeListener()中我们通过 checkedId 与 RadioButton 的 Id 作对比，如 checkedId==rbSex1.getId()则选择的是第一个 RadioButton。对 CheckBox 的事件监听 setOnCheckedChangeListener()中通过 isChecked()来判断选项是否被选中。

在主程序中，我们还需要对"提交"按钮来设置监听。程序中通过 btnOK.setOnClickListener() 来设置监听，首先要判断用户是否填写所有选项，然后通过意图（Intent）来将用户的信息传递到下一个 Activity 中。我们首先创建一个 Intent，并通过 intent.setClass(CH3_4.this, CH3_4_01.class) 为其指定它的类，然后我们需要创建一个绑定（Bundle）对象，并通过 bundle.putString("name", userName)来将第 1 个 Activity 获得的数据已变量的形式传递到下一个 Activity 中，通过 intent.putExtras(bundle);将 bundle 赋给所创建的 intent 中。最后通过 startActivity(intent)启动第 2 个 Activity 并将 intent 传递过去。在第 2 个 Activity 中，我们只需要解析出 Intent 中传递过来的数据，然后通过 setText()设置到 TextView()中，详细代码参考本书所附的源代码。

在完成了上述的程序后，也许你会迫不及待地想测试一下自己的程序，这时你运行的结果也许会让你失望，我们还有关键的一个步骤。因为在我们的例子中有两个 Activity，所以我们需要在 AndroidManifest.xml 里必须具备这两个 activity 的声明。本程序 AndroidManifest.xml 源码如代码清单 3-12 所示。

代码清单 3-12 第 3 章\CH3_4\AndroidManifest.xml

```xml
<?xml version="1.0" encoding="utf-8"?>
<manifest xmlns:android="http://schemas.android.com/apk/res/android"
      package="sziit.practice.chapter3"
      android:versionCode="1"
      android:versionName="1.0">
    <application android:icon="@drawable/icon" android:label="@string/app_name">
        <activity android:name=".CH3_4"
                  android:label="@string/app_name">
            <intent-filter>
                <action android:name="android.intent.action.MAIN" />
                <category android:name="android.intent.category.LAUNCHER" />
            </intent-filter>
```

```
    </activity>
    <activity android:name="CH3_4_01"></activity>
</application>
<uses-sdk android:minSdkVersion="8" />
</manifest>
```

3.3.5　知识点 5：数字时钟与模拟时钟（AnalogClock，AnalogClock）

我们在软件开发中有时候会用到钟表等控件来显示当前时间，同样地，在手机开发中也一样，在 Android 开发的 SDK 中，为我们提供了一个方便的钟表控件使用方法，让我们无需做任何编码就可在手机界面中显示时钟状的钟表 UI 图像，让界面看起来更加生动活泼。

Android 中的 AnalogClock Widget 是一个时钟控件类，我们在范例 CH3_05 中通过配置一个小的模拟时钟并在其下放一个 TextView 来显示数字信息，并在最后加上一个数字时钟以做对比。在这个例子中，会用到 Android.os.Handler、java.lang.Thread 和 Android.os.Message 对象的应用。通过 Thread 对象，在进程内同步调用 System.currentTimeMillis()获取系统时间，并通过 Message 对象来通知 Handler 对象，Handler 则扮演联系 Activity 与 Thread 之间的桥梁，在收到 Message 对象后，将时间变量的值显示到模拟时钟下面的 TextView 中，产生与数字时钟相同的效果。

程序运行结果如图 3-13 所示。

图 3-13　数字时钟（DigitalClock）与模拟时钟（AnalogClock）

程序布局文件代码如代码清单 3-13 所示。

代码清单 3-13　第 3 章\CH3_5\res\layout\main.xml

```xml
<?xml version="1.0" encoding="utf-8"?>
<LinearLayout xmlns:android="http://schemas.android.com/apk/res/android"
    android:orientation="vertical"
    android:layout_width="fill_parent"
    android:layout_height="fill_parent"
    >
<AnalogClock
    android:id="@+id/analogClock"
    android:layout_width="wrap_content"
    android:layout_height="wrap_content"
    />
<TextView
    android:id="@+id/digitalClock"
    android:layout_width="wrap_content"
    android:layout_height="wrap_content"
    />
<DigitalClock
    android:layout_width="wrap_content"
    android:layout_height="wrap_content"
    />
</LinearLayout>
```

当然我们可以在布局文件中设置模拟时钟和数字时钟的属性。数字时钟和模拟时钟所对应的标签为<DigitalClock>和<AnalogClock>。布局文件相对比较简单，下面就来了解在主程序中的实现，如代码清单 3-14 所示。

代码清单 3-14　第 3 章\ CH3_5\src\sziit\practice\chapter3 \CH3_5.java

```java
//包引入语句省略
public class CH3_5 extends Activity {//从基类Activity派生子类CH3_5
    protected static final int GUINOTIFIER=0x1234;　//声明一常数作为判别信息用
    //声明TextView对象和AnalogClock对象变量
    private TextView tv；//定义私有TextView对象
    private AnalogClock aClock；//定义私有AnalogClock对象
    //声明与时间有关的变量
    public Calendar calendar；//定义公有Calendar对象
    public int minutes;
    public int hours;
```

```java
//声明关键Handler和Thread变量
public Handler handler；  //定义公有Handler对象
private Thread thread；  //定义私有Thread对象
public void onCreate(Bundle savedInstanceState) {//程序加载时首先被调用
    super.onCreate(savedInstanceState)；  //调用基类onCreate方法
    setContentView(R.layout.main)；  //设置屏幕布局，通过R.layout.main引用布局资源
    //获取TextView对象和AnalogClock对象
    tv=(TextView)findViewById(R.id.digitalClock)；  //引用布局文件中定义的TextView
    aClock=(AnalogClock)findViewById(R.id.analogClock)；  //引用AnalogClock对象
    handler=new Handler(){//创建Handler对象
        public void handleMessage(Message msg){//子类重写handleMessage处理消息循环
            switch(msg.what){//识别具体的消息
                case CH3_5.GUINOTIFIER:
                    tv.setText(hours+":"+minutes);
                    break;
            }
            super.handleMessage(msg)；  //其他消息交给基类处理
        }
    };
    //通过进程获取系统时间
    thread=new LooperThread();
    thread.start()；  //启动进程
}
class LooperThread extends Thread{//从基类Thread派生子类LooperThread
    public void run(){//子类重写基类run方法
        super.run()；  //调用基类run方法
        try{//防止程序出现异常
            do{
                long time=System.currentTimeMillis()；  //获取系统时间
                calendar=Calendar.getInstance()；  //获取Calendar对象实例
                calendar.setTimeInMillis(time)；  //设置Calendar对象信息
                hours=calendar.get(Calendar.HOUR)；  //获得小时
                minutes=calendar.get(Calendar.MINUTE)；  //获取分钟
                Thread.sleep(1000)；  //每隔1s更新一次
                Message m=new Message()；  //创建Message对象
                m.what=CH3_5.GUINOTIFIER;
                CH3_5.this.handler.sendMessage(m)；  //发送消息
```

}**while**(CH3_5.LooperThread.*interrupted*()==**false**)；//系统发出中断信息时停止本循环

　　　　　}**catch**(Exception e){//捕获异常
　　　　　　　e.printStackTrace();　//打印异常信息
　　　　　}
　　　}
　　}
}

　　在主程序中我们需要另外加载 Java 的 Calendar 与 Thread 对象，在 onCreate()中构造 Handler 与 Thread 两个对象，并实现 handleMessage()和 run()方法。通过建立进程 LooperThread 来持续取得系统的时间，当我们取得系统的时间后则通过 Message 来通知 Handler 更新 TextView 中的数字。在设置发送信息的内容的时候已设置了一个常量 GUINOTIFER 来作为判别信息用，当获取系统时间成功后，发送给 Handler 的信息内容通过 m.what=CH3_5.GUINOTIFIER 来设置成所声明的常量。在 Handler 中我们根据 Message 的内容通过 switch（m.what）来处理 TextView 数字的更新。

　　以上实现了用 TextView 来显示数字时钟的信息，最后我们可以和<DigitalClock>的显示效果做比较。

　　在 Android 中提供了 System.currentTimeMillis()、uptimeMillis()和 elapsedRealtime()3 种不同的 SystemClock 给用户开发使用。第一种是标准的 Clock 用法，在实例中就是使用了这种方法，需要搭配真实的日期与时间使用。另外两种适用于 interval 与 elapse time 来控制程序与 UI，读者可自己动手练习一下。

3.3.6　知识点 6：日期与时间（DatePicker，TimePicker）

　　时间和日期是在日常的应用中最常见的基本应用，无论是在日常的网络服务还是在手机上的基本应用，时间和日期都是任何手机都有的基本功能。在 Android 中，同样有两个组件来实现日期和时间的功能，那就是 DatePicker 和 TimePacker。DatePicker 用来实现日期，TimePicker 用来实现时间。这里需要注意：在 Android 中还提供了另外两种其他的对象来实现动态修改日期时间的功能：DatePickerDialog 与 TimePickerDialog。这两种类型的对象最大的差别在于 DatePicker 与 TimePicker 是直接显示在屏幕上的，而 DatePickerDialog 和 TimePickerDialog 则是以对话框的形式显示在屏幕上。下面我们就通过一个实例来看一下这两组不同组件的应用。

　　在本节的实例（源码详见：第 3 章\CH3_6）中，我们使用 DatePicker、DatePickerDialog、TimePicker、TimePickerDialog、TextView 以及 Button 这 6 种常用的组件来实现实例。我们用 TextView 来显示日期和时间，默认为系统当前的时间，然后通过 Button 来调用 DatePickerDialog 与 TimePickerDialog 进行对日期和时间的修改，并动态显示到 TextView 中。

　　程序运行结果如图 3-14 所示。

　　单击"设置日期"或者"设置时间"就会调用按钮的事件监听，调出 DatePickerDialog 或者 TimePickerDialog，如图 3-15 所示。

第 3 章 Android 控件 Widgets

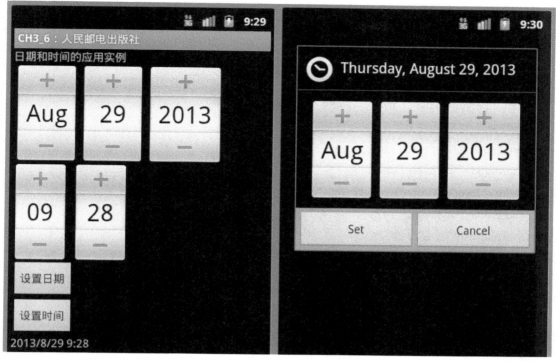

图 3-14　DatePicker 和 TimePicker　　　　　　图 3-15　DatePickerDialog

根据图 3-15 所示，我们可以看到布局很简单，由 DatePicker、TimePicker、两个 TextView 以及两个 Button 组成。布局文件如代码清单 3-15 所示。

代码清单 3-15　第 3 章\CH3_6\res\layout\main.xml

```xml
<?xml version="1.0" encoding="utf-8"?>
<LinearLayout xmlns:android="http://schemas.android.com/apk/res/android"
    android:orientation="vertical"
    android:layout_width="fill_parent"
    android:layout_height="fill_parent"
    >
<TextView
    android:layout_width="fill_parent"
    android:layout_height="wrap_content"
    android:text="日期和时间的应用实例"
    />
<DatePicker
    android:id="@+id/datePicker"
    android:layout_width="wrap_content"
    android:layout_height="wrap_content"
```

```
    />
<TimePicker
    android:id="@+id/timePicker"
    android:layout_width="wrap_content"
    android:layout_height="wrap_content"
    />
<Button
    android:id="@+id/btnDateSet"
    android:layout_width="wrap_content"
    android:layout_height="wrap_content"
    android:text="设置日期"/>
<Button
    android:id="@+id/btnTimeSet"
    android:layout_width="wrap_content"
    android:layout_height="wrap_content"
    android:text="设置时间"/>
<TextView
    android:id="@+id/tv"
    android:layout_width="wrap_content"
    android:layout_height="wrap_content"/>
</LinearLayout>
```

从上面的布局文件中我们可以知道，对 DatePicker 和 TimePicker 的调用，可以直接在布局文件中使用标签<DatePicker>和<TimePicker>来使用组件，当然也可以在布局文件中设置它们的属性。

接下来我们就要从主程序中来了解程序的具体实现。通过 updateShow()方法来更新 TextView 中所显示的日期和时间，并通过 java.util.Calendar 对象来取得当前的系统时间，这个方法我们在上一节已经讲到过了。当用户更改了 DatePicker 中的日期信息和 TimePicker 中的系统时间时则调用 onDateChanged()和 onTimeChanged()事件，并运行 updateShow()来重新加载 TextView 中显示的时间。除此之外，我们还设置两个按钮：“设置日期”和“设置时间”。当用户单击两个按钮时就会触发监听事件，从而调用 DatePickerDialog 与 TimePickerDialog 进行对日期和时间的修改。代码清单 3-16 所示的是主程序的实现部分。

代码清单 3-16 第 3 章\CH3_6\src\sziit\practice\chapter3\CH3_6.java

```java
public class CH3_6 extends Activity {//从Activity基类派生子类CH3_6
    /*声明日期相关的变量*/
    private int intYear; //定义私有变量，保存"年"信息
    private int intMonth; //定义私有变量，保存"月"信息
    private int intDay; // 定义私有变量，保存"天"信息
```

```java
/*声明时间相关的变量*/
private int intHour;  //定义私有变量,保存"小时"数据
private int intMinute;  //定义私有变量,保存"分钟"数据
/*声明对象变量*/
TextView tv;  //定义TextView对象
DatePicker dpDate;  //定义DatePicker对象
TimePicker tpTime;  //定义TimePicker对象
//声明Calendar类的对象
Calendar c;
//声明按钮对象
Button btnSetDate;
Button btnSetTime;
public void onCreate(Bundle savedInstanceState) {//子类重写基类onCreate方法
    super.onCreate(savedInstanceState);  //调用基类onCreate方法
    setContentView(R.layout.main);  //设置屏幕布局
    //获得系统的当前时间与日期,将获得的时间和日期赋给相应的日期和时间变量
    c=Calendar.getInstance();  //获得日历Calendar对象实例
    intYear=c.get(Calendar.YEAR);  //提取"年"数据
    intMonth=c.get(Calendar.MONTH);  //提取"月"数据
    intDay=c.get(Calendar.DAY_OF_MONTH);  //提取"天"数据
    intHour=c.get(Calendar.HOUR_OF_DAY);  //提取"小时"数据
    intMinute=c.get(Calendar.MINUTE);  //提取"分钟"数据
    //获取TextView对象
    tv=(TextView)findViewById(R.id.tv);  //从布局文件中,通过R.id.tv查找TextView对象
    updateShow();  //更新显示信息
    //获取DatePicker对象,并将DatePicker对象初始化为当前的系统时间,并设置事件监听
    dpDate=(DatePicker)findViewById(R.id.datePicker);  //引用布局文件中DatePicker对象
    dpDate.init(intYear, intMonth, intDay, new DatePicker.OnDateChangedListener(){//初始化
        public void onDateChanged(DatePicker view, int year,
                int monthOfYear, int dayOfMonth) {//处理日期变换事件响应
            intYear=year;
            intMonth=monthOfYear;
            intDay=dayOfMonth;
            //调用updateShow()更新TextView中的日期
            updateShow();
        }
    });
```

```java
        //获取TimePicker对象,并设置时间显示的形式为24小时制
        tpTime=(TimePicker)findViewById(R.id.timePicker);
        tpTime.setIs24HourView(true);
        //设置事件监听,注意与DatePicker中事件监听的区别
        tpTime.setOnTimeChangedListener(new TimePicker.OnTimeChangedListener(){
            public void onTimeChanged(TimePicker view, int hourOfDay, int minute) {
                intHour=hourOfDay;
                intMinute=minute;
                updateShow();
            }
        });
        //获取设置时间和日期的两个按钮对象,并设置事件监听
        btnSetDate=(Button)findViewById(R.id.btnDateSet);
        btnSetTime=(Button)findViewById(R.id.btnTimeSet);
        btnSetDate.setOnClickListener(new Button.OnClickListener(){//设置单击监听事件
            public void onClick(View arg0) {//处理单击事件
                new DatePickerDialog(CH4_10.this,intDay, new
DatePickerDialog.OnDateSetListener(){//创建DatePickerDialog对象
                    public void onDateSet(DatePicker view, int year,
                            int monthOfYear, int dayOfMonth) {
                        //更新日期的方法
                    }
                }, intYear, intMonth, intDay).show();  //显示DatePickerDialog对话框
            }
        });
        btnSetTime.setOnClickListener(new Button.OnClickListener(){//设置单击监听事件
            public void onClick(View v) {//设置按钮单击事件响应
                new TimePickerDialog(CH4_10.this,intDay, new
TimePickerDialog.OnTimeSetListener(){//创建TimePickerDialog对象
                    public void onTimeSet(TimePicker view, int hourOfDay,
                            int minute) {//重写onTimeSet方法
                        //更新时间的方法
                    }
                }, intHour, intMinute, true).show();  //显示TimePickerDialog对话框
            }
        });
    }
```

```
protected void updateShow() {//更新显示信息
    tv.setText(new StringBuilder().append(intYear).append("/").append(intMonth+1)
            .append("/")
            .append(intDay).append(" ")
            .append(intHour).append(":")
            .append(intMinute));
    }
}
```

在这个实例中我们重点学习如何使用 DatePicker 和 TimePicker 对象来动态地调整日期和时间。在上面的程序中我们需要注意：DatePicker 实现 OnDateChangedListener()的方法与 TimePicker 实现 OnTimeChangedListener()的方法的不同之处。DatePicker 对象以 init()这个方法来指定 DatePicker 初始的年、月、日以及 OnDateChangedListener()的事件；而 TimePicker 对象则是直接以 setOnTimeChangedListener()事件来处理时间改变时程序要做的操作。

对于 DatePickerDialog 与 TimePickerDialog 的实现，我们则是在按钮的监听事件中通过新建 DatePickerDialog 对象来实现的，如下所示：

new DatePickerDialog (Context context, int theme, DatePickerDialog.OnDateSetListener callBack, int year, int monthOfYear, int dayOfMonth).show()

注意最后的.show()方法的作用是将所建的 DatePickerDialog 对话框显示出来。同样的道理，TimePickerDialog 的实现也类似。读者可以对照代码清单 3-16 来学习。

最后，我们用 updateShow()来更新 TextView 中的信息，通过 TextView 对象的 setText()来实现。

3.4 项目实施

在本章中我们学习了 Android 平台中基本控件的使用。Android 为我们提供了多样化的可以灵活使用的控件。如何能够合理的使用需要我们在开发中不断地摸索和总结。我们对控件的使用通常要与我们之前学过的布局相结合。那么下面我们就通过一个简单的项目来实际应用我们本章所学的内容。效果如图 3-16 所示。

此项目需要完成的功能是按照用户的选项进行与后台的服务器进行交互，获得用户所选条件下的数据信息。布局和各个控件的代码清单 3-17 如下所示。

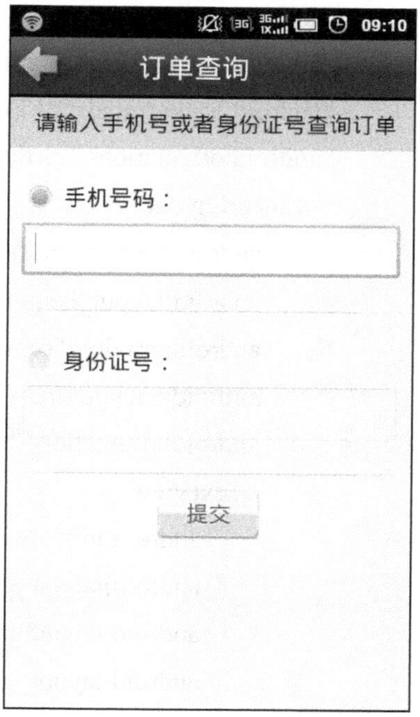

图 3-16 项目实施效果

代码清单 3-17

```xml
<?xml version="1.0" encoding="UTF-8"?>
<LinearLayout xmlns:android="http://schemas.android.com/apk/res/android"
    android:layout_width="match_parent"
    android:layout_height="match_parent"
    android:background="#FFF4F9F9"
    android:orientation="vertical" >
<LinearLayout
        android:layout_width="fill_parent"
        android:layout_height="wrap_content"
        android:background="@drawable/title_background"
        android:gravity="center"
        android:orientation="horizontal" >
        <include layout="@layout/comm_title_bar" />
</LinearLayout>
<ScrollView android:layout_width="fill_parent"
            android:layout_height="fill_parent">
    <LinearLayout
        android:layout_width="fill_parent"
        android:layout_height="fill_parent"
        android:background="#FFF4F9F9"
        android:orientation="vertical" >
        <LinearLayout
            android:layout_width="fill_parent"
            android:layout_height="wrap_content"
            android:gravity="center"
            android:background="@drawable/order_manage_back"
            android:orientation="horizontal" >
            <TextView
                android:id="@+id/reminde_info"
                android:layout_width="wrap_content"
                android:layout_height="wrap_content"
                android:layout_gravity="center_vertical"
                android:layout_marginBottom="5dip"
                android:layout_marginLeft="5dip"
                android:layout_marginTop="5dip"
                android:text="@string/order_reminder"
```

```xml
            android:textAppearance="?android:textAppearanceMedium"
            android:textColor="@color/black" />
</LinearLayout>
<LinearLayout
        android:layout_width="fill_parent"
        android:layout_height="wrap_content"
        android:background="#FFF4F9F9"
        android:gravity="center_vertical"
        android:orientation="horizontal" >
    <RadioGroup
            android:id="@+id/myRadioGroup"
            android:layout_width="fill_parent"
            android:layout_height="wrap_content"
            android:orientation="vertical">
        <RadioButton
            android:id="@+id/radio_telephone"
            android:layout_width="wrap_content"
            android:layout_height="wrap_content"
            android:layout_marginLeft="5dip"
            android:layout_marginTop="20dip"
            android:text="@string/telephone"
            android:textColor="#FF000000"
            android:textSize="18dip" />
        <EditText
            android:id="@+id/hotel_manage_telephone"
            android:layout_width="fill_parent"
            android:layout_height="wrap_content"
            android:layout_weight="1"
            android:inputType="number"
            android:maxLength="18"
            android:layout_marginLeft="10dip"
            android:layout_marginRight="10dip"
            android:layout_marginBottom="20dip"
            android:textSize="18dip" />
        <ImageView
            android:layout_width="fill_parent"
            android:layout_height="5dip"
```

```xml
            android:src="@drawable/divide" />
        <RadioButton
            android:id="@+id/radio_iden"
             android:layout_width="wrap_content"
            android:layout_height="wrap_content"
            android:layout_marginLeft="5dip"
            android:text="@string/iden"
            android:layout_marginTop="20dip"
            android:textColor="#FF000000"
            android:textSize="18dip" />
        <EditText
            android:id="@+id/hotel_manage_iden"
            android:layout_width="fill_parent"
            android:layout_height="wrap_content"
            android:layout_weight="1"
            android:maxLength="18"
             android:layout_marginLeft="10dip"
            android:layout_marginRight="10dip"
              android:layout_marginBottom="20dip"
            android:textSize="18dip" />
    </RadioGroup>
  </LinearLayout>
<ImageView
      android:layout_width="fill_parent"
      android:layout_height="5dip"
      android:layout_marginBottom="5dip"
      android:src="@drawable/divide" />
<LinearLayout
      android:layout_width="wrap_content"
      android:layout_height="wrap_content"
      android:background="#FFF4F9F9"
      android:layout_gravity="center_horizontal" >
        <Button
        android:id="@+id/hotel_query_button"
        android:layout_width="80dip"
        android:layout_height="40dip"
        android:layout_margin="5dip"
```

```
            android:background="@drawable/order_btn_bg"
            android:padding="5dip"
            android:text="@string/commit"
            android:textSize="18dip" />
        </LinearLayout>
    </LinearLayout>
  </ScrollView>
</LinearLayout>
```

3.5 技术拓展

上一节我们简单介绍了用户界面的知识，用户界面除了显示给用户观看外还需要响应用户的各种操作的事件。什么是事件呢？所谓事件，就是用户与 UI（图形界面）交互时所触发的操作。例如，在手机键盘上按下一个键时就触发了"按下"事件，当松开时又触发了"弹起"事件。Android 平台使用回调机制来处理用户界面事件，每一个 View 都有自己的处理事件的回调方法，如果事件没有被 Activity 的任何一个 View 所处理时，Android 就会调用 Activity 的事件处理回调方法进行处理。下面我们简单介绍一些常用的事件处理的回调方法。

- public boolean onKeyDown(int keycode,KeyEvent event)：处理手机上按键被按下事件的回调方法。
- public boolean onKeyUp(int keycode,KeyEvent event)：处理手机上按键按下后弹起事件的回调方法。
- public boolean onTouchEvent(MotionEvent event)：处理触摸事件的回调方法。
- public boolean onKeyMultiple(int keycode,int repeatCount,KeyEvent event)：按键重复单击时的回调方法。
- Protected void onFocusChanged(boolean gainFocus,int direction,Rect previous)：焦点改变时的回调方法。

在 Android 内置的控件类中，根据功能的需要，已经实现了一些回调方法来处理事件。当我们需要更改或者添加对事件处理的行为时，则需要继承这些控件类并重新实现自己的事件处理方法。

在上面的回调方法中我们提到了焦点问题。焦点描述了按键事件（包括轨迹球）的承受者，即每个按键事件都是发生在当前拥有焦点的 View 上。焦点可以移动到 View 上，View 也可以设置移动的下一个 View 的 id。父控件，还可以阻止子控件获得焦点。表 3-3 中我们列出了与 View 的焦点有关的方法。

表 3-3　　　　　　　　　　　与 View 的焦点有关的方法

方　　法	简　　介
setFocusable()	设置 View 是否可以拥有焦点
isFocusable()	返回 View 是否已经拥有焦点

续表

方　　法	简　　介
setFocusableInTouchMode()	设置 View 是否可在触摸模式获得焦点，默认 false
isFocusableInTouchMode()	返回 View 是否在触摸模式已经获得了焦点
requestFocus()	尝试让 View 获得焦点
isFocused()	返回 View 当前是否获得了焦点
hasFocus()	返回 View 的父控件是否获得了焦点
hasFocusable()	返回 View 的父控件是否允许当前 View 获得焦点，或者当前 View 是否有可以获得焦点的子控件
setNextFocusDownId() setNextFocusLeftId() setNextFoucsRightId() setNextFoucsUpId()	分别设置 View 的焦点在下、左、右、上移动后获得焦点的 id

　　除了重新实现上述的这些回调方法，Android 还为每个 View 提供了一种监听事件的接口，每个接口都需要实现一个回调方法，然后使用事件监听接口。使用这种监听接口的好处是，当不同的 View 触发了相同类型的事件时可以调用同一个回调方法，并且也不必为了处理事件而重新定义自己的控件类了。下面我们简单介绍这些事件监听接口。

- OnClickListener，单击（Click）事件的监听接口。在触摸模式中，单击事件是指在 View 上按下并在 View 上抬起的组合动作，就像在计算机上单击按钮一样。通过按键则是在 View 获得焦点时按下轨迹球或者"返回"键以触发单击事件。

    回调方法：public void onClick(View v)

- OnLongClickListener，长按事件的监听接口。在触摸模式中，长按事件是指在 View 上长按住不放。通过按键则是在 View 获得焦点时按住轨迹球或者"返回"键保持约 1s 以上来触发事件。

    回调方法：public boolean onLongClick(View v)

- OnFocusChangeListener，焦点变化事件的监听接口。当通过轨迹球使 View 获得或者失去焦点时触发焦点变化事件。

    回调方法：public void onFocusChange(View v,boolean hasFocus)

- OnKeyListener，按键事件的监听接口。当 View 获得焦点时，按下或者抬起手机上的任意按键触发按键事件。

    回调方法：public boolean onKey(View v,int keyCode,KeyEvent event)

- OnTouchListener，触摸事件的监听接口。当用户在 View 界面范围内触摸按下、抬起或者滑动的动作时触发触摸事件。

    回调方法：public boolean onTouch(View v,MotionEvent event)

- OnCreateContextMenuListener，上下文菜单显示事件监听接口。当 View 使用 showContextMenu()时，触发上下文菜单显示事件。

  回调方法：public void onCreateContextMenu(ContextMenu menu, View v, ContextMenuInfo info)

上面我们介绍了 Android 中事件处理的基本知识，接下来通过一个例子来演示一下 Android 事件处理的方法。实例参见本书源代码：第 3 章\CH3_7。程序运行后的结果如图 3-17～图 3-20 所示。其中，图 3-17 表示按下"范例按钮"，图 3-18 表示按下向右的方向键，图 3-19 表示的是按下向下的方向键后弹起的效果图，图 3-20 表示的是获取触笔在屏幕上的坐标。

图 3-17　按下范例按钮

图 3-18　按下向右方向键

图 3-19　弹起向下方向键

图 3-20　捕获触笔坐标

从上面程序运行的截图我们可以看到，在这个程序中用户界面只有一个 TextView 和一个 Button。请读者自己试着布局一下这个程序的用户界面。下面我们给出这个界面的布局文件的源代码，读者可以对照一下。

代码清单 3-18　第 3 章\CH3_7\res\layout\main.xml

```xml
<?xml version="1.0" encoding="utf-8"?>
<LinearLayout xmlns:android="http://schemas.android.com/apk/res/android"
    android:orientation="vertical"
    android:layout_width="fill_parent"
    android:layout_height="fill_parent"
    >
<TextView
    android:layout_width="fill_parent"
    android:layout_height="wrap_content"
    android:text="@string/hello"
    />
<Button
    android:id="@+id/btnTest"
    android:layout_width="wrap_content"
    android:layout_height="wrap_content"
    android:text="范例按钮"/>
</LinearLayout>
```

在布局完成后就来看一下程序的主要部分。程序代码如代码清单 3-19 所示。

代码清单 3-19　第 3 章\CH3_7\src\sziit\practice\chapter3\CH3_7

```java
//包的引入语句省略
public class CH3_7 extends Activity {//从Activity基类派生子类CH3_7
    public void onCreate(Bundle savedInstanceState) {//子类重写基类onCreate方法
        super.onCreate(savedInstanceState);    //调用基类onCreate方法
        setContentView(R.layout.main);    //设置屏幕布局，通过R.layout.main引用布局资源
        Button btnTest=(Button)findViewById(R.id.btnTest);    //通过findViewById查找Button对象
        btnTest.setOnClickListener(new Button.OnClickListener(){//单击（Click）事件的监听接口
            public void onClick(View v){//处理onClick回调方法
                ShowMessage("您单击了范例按钮");    //显示处理信息
            }
        });
    }
    //按下按键时触发的事件
    public boolean onKeyDown(int keyCode,KeyEvent event){//重写onKeyDown回调方法
```

```java
switch(keyCode){//处理按下的按钮
case KeyEvent.KEYCODE_DPAD_CENTER:
    ShowMessage("按下中键");
    break;
case KeyEvent.KEYCODE_DPAD_DOWN:
    ShowMessage("按下了向下方向键");
    break;
case KeyEvent.KEYCODE_DPAD_LEFT:
    ShowMessage("按下了向左方向键");
    break;
case KeyEvent.KEYCODE_DPAD_RIGHT:
    ShowMessage("按下了向右方向键");
    break;
case KeyEvent.KEYCODE_DPAD_UP:
    ShowMessage("按下了向上方向键");
    break;
}
return super.onKeyDown(keyCode, event);  //其他按键交给基类的onKeyDown方法处理
}
//按键弹起时触发的事件
public boolean onKeyUp(int keyCode,KeyEvent event){//处理onKeyUp回调方法
switch(keyCode){//处理释放的按键
case KeyEvent.KEYCODE_DPAD_CENTER:
    ShowMessage("弹起中键");
    break;
case KeyEvent.KEYCODE_DPAD_DOWN:
    ShowMessage("弹起向下方向键");
    break;
case KeyEvent.KEYCODE_DPAD_LEFT:
    ShowMessage("弹起向左方向键");
    break;
case KeyEvent.KEYCODE_DPAD_RIGHT:
    ShowMessage("弹起向右方向键");
    break;
case KeyEvent.KEYCODE_DPAD_UP:
    ShowMessage("弹起向上方向键");
    break;
```

}
　　return super.onKeyUp(keyCode, event);　//其他按键交给基类的onKeyUp方法处理
}
//捕获触笔在屏幕上的坐标
public boolean onTouchEvent(MotionEvent event){//处理回调方法onTouchEvent
　　int iAction=event.getAction();
　　if(iAction==MotionEvent.ACTION_CANCEL||iAction==MotionEvent.ACTION_DOWN||iAction==MotionEvent.ACTION_MOVE){
　　　　return false;
　　}
　　int x=(int)event.getX();
　　int y=(int)event.getY();
　　ShowMessage("触笔坐标：（"+Integer.toString(x)+","+Integer.toString(y)+")");
　　return super.onTouchEvent(event);　//其他消息交给基类的onTouchEvent方法处理
}
public void ShowMessage(String message){//显示消息
　　Toast.makeText(this, message, Toast.LENGTH_SHORT).show();　//调用Toast显示提示信息
}
}
```

在上面的代码中，通过设置控件的监听器来监听并重写了某些方法来实现事件处理。我们通过定义方法 ShowMessage()来向用户进行提示。注意：在 ShowMessage()方法中我们用到了 Toast.makeText(this,string,Toast.LENGTH_SHORT).show();来显示一个短时间的提示信息。关于 Toast 的指示我们在后面会进行介绍。

## 3.6　本章小结

在本章我们结合上一章的布局知识，主要讲解了 Android 的基本控件的知识。通过本章的学习，相信大家对 Android 中的基本控件的使用以及使用方法已有了大体的了解。同时对事件的交互也有了初步的理解。

然而 Android 为我们提供的不仅仅是以上所介绍的控件，更多的控件需要大家在日常的学习中积累、自定义。只有掌握了更多控件的使用方法，才能让我们的程序在功能的实现上有更多、更好的选择。

## 3.7　强化练习

一、填空题与简答题

1. Android 系统为我们提供了与用户进行交互的文本框信息，其名称为_____。
2. 设置 TextView 中文本的字体颜色有两种方式，分别为_____和_____。
```

3．如何设置按钮的事件监听？

4．介绍 RadioButton 与 RadioGroup 的关系。

二、编程题

在本章我们学习了 Android 的基本控件，包括 Button、EditText、TextView 等。那么下面就通过一个具体的实例来锻炼我们对控件的使用。我们需要完成下面一个简单的实例。完成如图 3-21 和图 3-22 所示的布局，并能够实现如下功能。

能够单击对应的按钮展开和收起信息：初始状态如图 3-21 所示，我们通过单击"门票信息"就能够把门票信息的优惠情况展示出来，同时按钮上的"箭头"也发生变化。如图 3-22 所示。

图 3-21　初始状点

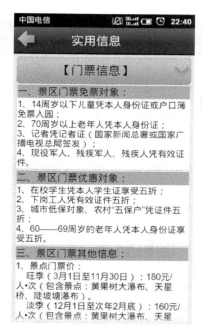

图 3-22　展示门票信息

第4章 Android 的图形用户界面

4.1 项目导引

前面一章我们学习了 Android 基本控件的使用，并且能够结合布局的知识创建一个具有良好用户体验风格的界面。但是，仅仅只有这些是远远不够的。当我们需要与后台进行访问以及数据加载时进度条又该如何实现呢？当用户的操作有问题需要给用户进行提示或者警告时又该如何操作呢？

上面的这些问题都是这一章我们需要学习和掌握的。

4.2 项目分析

为了能够让我们的软件具有更好的用户体验，Android 提供了在 PC 上经常见到和使用的对话框，通过对话框让用户与软件进行交流；为了避免让用户在使用软件以及在软件加载数据时不知所措，Android 提供了进度条，让用户知道我们的软件是在运转的，如图 4-1 所示。

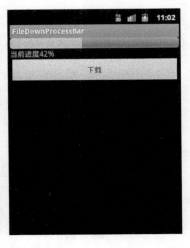

图 4-1 对话框与进度条

第 4 章 Android 的图形用户界面

在项目开发中经常会遇到列表，那么我们的 Android 有没有列表的控件呢？答案是肯定的。Android 不仅仅为开发者提供了列表，同样为开发者提供了"菜单"让我们使用。本章就从上面所讲的知识入手，带领大家进一步学习 Android 的基础知识。

4.3 技术准备

4.3.1 知识点 1：ListView

在 Android 中，ListView 是比较常用的控件，它以列表的形式展示具体内容，并且能够根据数据的长度进行自动适应并显示。用户可以选择并操作这个列表，同时触发一些事件。如当鼠标滚动时会触发 setOnItemSelectedListener 事件，单击时则会触发 setOnItemClickListener 事件。下面我们还是通过一个具体的实例来学习 ListView 的用法。

在实例 4-1 中我们将周一～周日放到 ListView 中，设置长按事件响应和选择事件响应，并在 TextView 中来显示所选日期。先来看下程序的运行结果，图 4-2 所示是当选择周二时，在 TextView 中便提示了所选择的日期。图 4-3 所示是用户在一个选项上长按后会弹出菜单的界面。

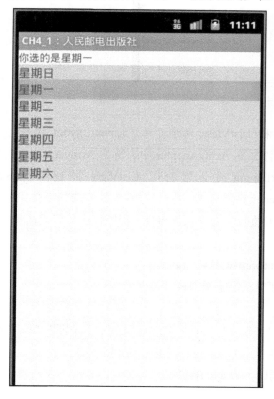

图 4-2　选择 ListView 中的选项

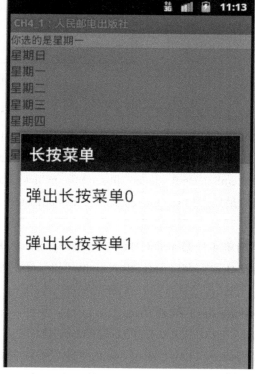

图 4-3　长按事件响应

还需要注意的是，在 ListView 中"选择"和"点击"是两个不同的概念。在例子中我们单击一个选项后会在 TextView 中提示"你点击的是……"，如图 4-4 所示。

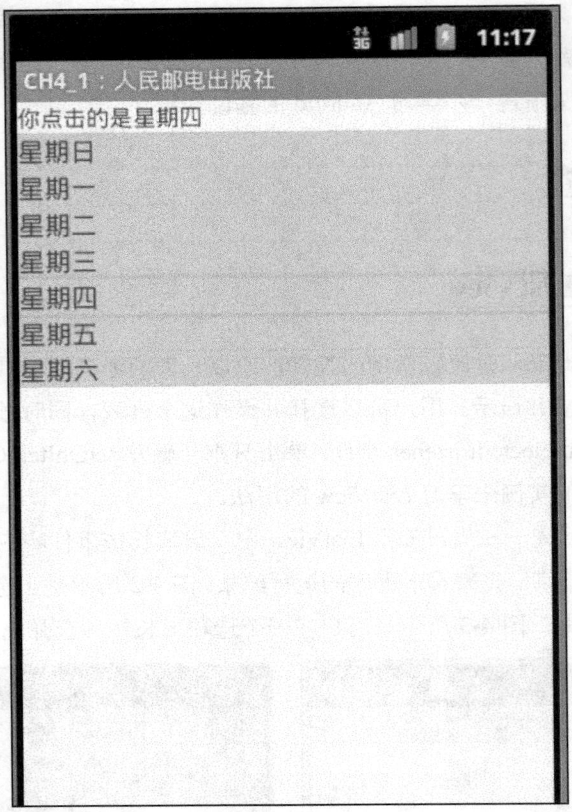

Android 应用程序设计

图 4-4 单击一个选项

在编码实现的时候需要创建 LinearLayout 和 ListView 两个对象。LinearLayout 用来显示 ListView，然后通过 ArrayAdapter 将获得的数组添加到 ArrayAdapter 中，再将 ArrayAdapter 添加到 ListView 中，将 ListView 添加到 LinearLayout 中，接着在 ListView 的视图中添加 setOnItemSelectedListener()和 setOnItemClickListener()两个监听方法，分别来处理选择事件和单击事件，最后添加 setOnCreateContextMenuListener()监听来处理长按事件。程序代码如代码清单 4-1 所示。

代码清单 4-1 第 4 章\CH4_1\src\sziit\practice\chapter4\CH4_1.java
public class CH4_1 **extends** Activity //从Activity基类派生子类CH4_1
{
 private static final String[] *array* = { "星期日", "星期一", "星期二", "星期三", "星期四", "星期五", "星期六"}; //定义私有静态只读String[]数组
 LinearLayout myLinearLayout; //定义线性布局LinearLayout对象
 TextView myTextView; //定义TextView对象
 ListView myListView; //定义ListView对象
 public void onCreate(Bundle savedInstanceState) //子类重写基类的onCreate方法
 {

```
super.onCreate(savedInstanceState);   //调用基类的onCreate方法
myLinearLayout = new LinearLayout(this);   //新增LinearLayout
myLinearLayout.setOrientation(LinearLayout.VERTICAL);   //设置成垂直线性布局方式
myLinearLayout.setBackgroundColor(android.graphics.Color.WHITE);   //将背景颜色设置为白色
myTextView = new TextView(this);   //新增TextView
LinearLayout.LayoutParams param1 =
    new LinearLayout.LayoutParams( LinearLayout.LayoutParams.FILL_PARENT,
LinearLayout.LayoutParams.WRAP_CONTENT);
myTextView.setText(R.string.title);   //设置myTextView对象显示的信息
myTextView.setTextColor(getResources().getColor(R.drawable.blue));   //设置字体颜色
myLinearLayout.addView(myTextView, param1);   //将TextView加到myLinearLayout
myListView = new ListView(this);   // 新增ListView
LinearLayout.LayoutParams param2 =
    new LinearLayout.LayoutParams( LinearLayout.LayoutParams.FILL_PARENT,
LinearLayout.LayoutParams.WRAP_CONTENT);
myListView.setBackgroundColor(getResources().getColor(R.drawable.ltgray));
myLinearLayout.addView(myListView, param2);   //将ListView加到myLinearLayout
setContentView(myLinearLayout);   //将LinearLayout加到ContentView
// new ArrayAdapter对象并将array字符串数组传入
ArrayAdapter adapter =
    new ArrayAdapter(this, R.layout.my_simple_list_item, array);
myListView.setAdapter(adapter);   //将ArrayAdapter加入ListView对象中
// myListView加入OnItemSelectedListener事件的监听器
myListView .setOnItemSelectedListener(new AdapterView.OnItemSelectedListener()
{
    public void onItemSelected(AdapterView arg0, View arg1, int arg2, long arg3)
    { //处理回调方法onItemSelected
        //使用getSelectedItem()将选取的值带入myTextView中
        myTextView.setText("你选的是" + arg0.getSelectedItem().toString());
    }
    public void onNothingSelected(AdapterView arg0)
    {
    }
});
// myListView加入OnItemClickListener事件的监听器
myListView.setOnItemClickListener(new AdapterView.OnItemClickListener()
{
```

```
    public void onItemClick(AdapterView arg0, View arg1, int arg2, long arg3)
    { //处理回调方法
        //使用String[index]，arg2为点选到ListView的index，并将值带入myTextView中
        myTextView.setText("你单击的是" + array[arg2]);
    }
});
//设置长按则弹出菜单
myListView.setOnCreateContextMenuListener(new OnCreateContextMenuListener() {
    public void onCreateContextMenu(ContextMenu menu, View v,
        ContextMenuInfo menuInfo) {
        menu.setHeaderTitle("长按菜单");
        menu.add(0, 0, 0, "弹出长按菜单0");
        menu.add(0, 1, 0, "弹出长按菜单1");
    }
});
  }
}
```

以上代码还用到了我们定义的 my_simple_list_item.xml，它定义了用来显示 ListView 中 Item 的组件，通过这个组件我们才可以获得用户所选择的 Item，其代码如代码清单 4-2 所示。

代码清单 4-2 第 4 章\CH4_1\res\layout\my_simple_list_item.xml

```xml
<?xml version="1.0" encoding="utf-8"?>
<CheckedTextView
    xmlns:android="http://schemas.android.com/apk/res/android"
    android:id="@+id/myCheckedTextView1"
    android:layout_width="fill_parent"
    android:layout_height="fill_parent"
    android:textColor="@drawable/black"
/>
```

此外还要注意，在程序中使用 LinearLayout 对象，动态地将 TextView 与 ListView 添加到原来 Layout 中的 main.xml 布局中。LinearLayout.LayoutParams param1：用来创建一个 LayoutParams 对象 param1，再调用 myLinearLayout.addView(myTextView,param1)来将 param1 对象传入，动态添加 TextView。myListView.setOnCreateContextMenuListener()用来添加上下文菜单选项，我们会在后面的章节中学习。

4.3.2　知识点 2：对话框（Dialog）

正如本节前面所讲的一样，对话框也是 Android 为我们提供的一种可见的组件。Android 提供

了不同类型的对话框,其实前面我们已经接触过对话框了,那就是 DatePickerDialog、TimePickerDialog。除了我们讲过的这两个对话框外,还有 Dialog,它就相当于 javascript 中的 Alert 一样,用来显示提示信息;还有一种是 AlertDialog,它是一个简单的可以交互的对话框,跟 JavaScript 中的 confirm 对话框是一样的效果:能够允许用户选择 yes 或者 no,设置单选按钮、复选按钮或者简单的文本输入。当然还有一种对话框就是 ProcessDialog。对于 ProgressDialog,在 Windows 窗口程序或者 Flash 程序中会经常见到,诸如"加载中……"等对话框。在 Android 系统中,我们则是通 ProgressDialog 来运行,这个类封装在 Android.app.ProgressDialog 里,但是需要注意一点:在 Android 中,ProgressDialog 必须要在后台程序运行完毕前,以 dismiss()方法来关闭取得焦点,否则程序就会陷入无法终止的无穷循环中;又或者在线程里不可有任何更改 Context 或 parent View 的任何状态、文字输出等事件,因为线程中的 Context 和 View 并不属于 parent,两者之间没有关联。

下面我们就通过一个例子来看一下关于 Android 中对话框的类型和使用方法。首先来模拟一个论坛发表评论时经常出现的一个场景,也就是当我们要发表评论时首先需要登录,登录后才可以发表评论。

在这个例子中,首先要为 Activity 设置布局,在这里使用的布局就是 Main.xml 文件,只是我们添加了一个 Button、一个 EditText 和一个 TextView。这里就不再详细列出布局文件的代码,只是将布局后的效果给读者,如图 4-5 所示,读者可以自己完成,源代码在本书光盘的第 4 章/CH4_2 中。

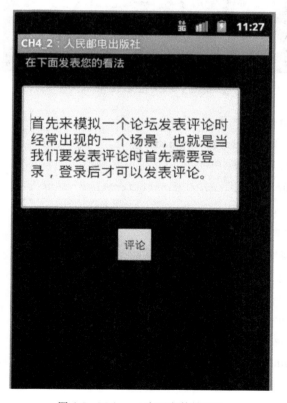

图 4-5 Main.xml 布局文件效果图

另外，在这个例子中还需要对我们自定义的对话框的布局进行设置。图 4-6 所示就是我们自己定义的登录框，其中包括两个 TextView 和两个 EditText。

图 4-6　自定义对话框

其布局文件如代码清单 4-3 所示。

代码清单 4-3　第 4 章\CH4_2\res\layout\login.xml

<?xml version="*1.0*" encoding="*utf-8*"?>
<LinearLayout xmlns:android="*http://schemas.android.com/apk/res/android*"
　　android:orientation="*vertical*"
　　android:layout_width="*fill_parent*"
　　android:layout_height="*wrap_content*"
　　>
<TextView
　　android:layout_height="*wrap_content*"
　　android:layout_width="*wrap_content*"
　　android:layout_marginLeft="*20dip*"
　　android:layout_marginRight="*20dip*"
　　android:text="*账号*"

```
        android:gravity="left"
        android:textAppearance="?android:attr/textAppearanceMedium"/>
<EditText
        android:id="@+id/userName"
        android:layout_height="wrap_content"
        android:layout_width="fill_parent"
        android:layout_marginLeft="20dip"
        android:layout_marginRight="20dip"
        android:scrollHorizontally="true"
        android:autoText="false"
        android:capitalize="none"
        android:gravity="fill_horizontal"
        android:textAppearance="?android:attr/textAppearanceMedium"/>
    <TextView
        android:layout_height="wrap_content"
        android:layout_width="wrap_content"
        android:layout_marginLeft="20dip"
        android:layout_marginRight="20dip"
        android:text="密码"
        android:gravity="left"
        android:textAppearance="?android:attr/textAppearanceMedium"/>
<EditText
        android:id="@+id/userName"
        android:layout_height="wrap_content"
        android:layout_width="fill_parent"
        android:layout_marginLeft="20dip"
        android:layout_marginRight="20dip"
        android:scrollHorizontally="true"
        android:autoText="false"
        android:capitalize="none"
        android:gravity="fill_horizontal"
        android:password="true"
        android:textAppearance="?android:attr/textAppearanceMedium"/>
</LinearLayout>
```

在上面的代码中需要注意几个属性的应用：android:gravity="left"表示对齐方向；android:textAppearance="?android:attr/textAppearanceMedium"表示文字的显示外观，其中有 4 个外观：

android:textAppearance="?android:attr/textAppearanceSmall"

android:textAppearance="?android:attr/textAppearanceMedium"

android:textAppearance="?android:attr/textAppearanceLarge"

android:textAppearance="?android:attr/textAppearanceLarge

代码中，android:scrollHorizontally="true" 表示当文字超出一行时是否出现滚动条；android:layout_marginLeft 和 android:layout_marginRight 表示左右边空白的大小。

在设置以上布局后，先来看一下主程序中的源代码，如代码清单 4-4 所示。

代码清单 4-4 第 4 章\CH4_2\src\sziit\practice\chapter4

```
//包的引入省略
public class CH4_2 extends Activity {//从 Activity 基类派生子类
    ProgressDialog logining; //定义 ProgressDialog 对象
    public void onCreate(Bundle savedInstanceState) {//子类重写基类的 onCreate 方法
        super.onCreate(savedInstanceState); //调用基类的 onCreate 方法
        setContentView(R.layout.main); //设置屏幕布局，通过 R.layout.main 引用布局资源
        Button btnCommont=(Button)findViewById(R.id.btnCom); //引用布局文件定义的 Button
        btnCommont.setOnClickListener(new OnClickListener(){//设置 OnClickListener 监听器
            public void onClick(View v) {//处理单击回调方法 onClick
                Dialog dialog=new AlertDialog.Builder(CH4_2.this)//创建 AlertDialog 对话框
                    .setTitle("登录提示")//设置登录提示
                    .setMessage("您还没有登录，请先登录")//设置提示信息
                    .setPositiveButton("确定",    //设置确定按钮
                        new DialogInterface.OnClickListener(){//设置 OnClickListener 监听器
                            public void onClick(DialogInterface dialog,
                                int which) {//处理单击回调方法 onClick
                                //跳转到登录框
                                LayoutInflater factory=LayoutInflater.from(CH4_2.this);
                                //用来将 xml 文件转成 View
                                final View DialogView=factory.inflate(R.layout.login, null);
                                AlertDialog dlg=new AlertDialog.Builder(CH4_13.this)
                                    .setTitle("登录框")
                                    .setView(DialogView)
                                    .setPositiveButton("确定", //设置确定按钮
                                        new DialogInterface.OnClickListener(){
                                            public void onClick(
                                                DialogInterface dialog,
                                                int which) {
                                                logining=ProgressDialog.show(CH4_2.this, "登录中", "正在
```

```
                            登录，请稍后……",true);
                    new Thread(){
                        public void run(){
                            try{
                                sleep(3000);
                            }catch(Exception e){
                                e.printStackTrace();
                            }
                            finally{
                                logining.dismiss();
                            }
                        }
                    }.start();
                }
            })
            .setNegativeButton("取消",//设置取消按钮事件
                    new DialogInterface.OnClickListener(){
                        public void onClick(DialogInterface dialog,
                                int which) {
                            CH4_2.this.finish();
                        }
                    }).create();
            dlg.show();
        }
    }).setNeutralButton("取消",
        new DialogInterface.OnClickListener(){
        public void onClick(DialogInterface dialog,int which){
        CH4_2.this.finish();
        }
    }).create();   //创建
        dialog.show();  //显示
    }});
    }
}
```

在例子中单击"评论"按钮后，弹出提示登录对话框，单击"确定"时就会跳转到登录框中，如图4-7所示。

图 4-7　登录示意图

在上面的代码中我们需要注意：

factory=LayoutInflater.from(CH4_2.this); //用来将 xml 文件转成 View

final View DialogView=factory.inflate(R.layout.login, null);

以上两行代码将设置的布局文件 login.xml 用 LayoutInflater 对象的 inflate 方法转换成 View。

另外，从上面的例子中我们可以看到用 AlertDialog 来创建对话框是非常方便的，使用 AlertDialog.Builder 可以很方便地创建指定内容及样式的对话框。AlertDialog.Builder 还提供了多种方法来定制对话框，大致介绍如下。

- setTitle(CharSequence title)，setTitle(int titleId)，设置标题字符串
- setSingleChoiceItems，设置为单选选项对话框
- setMultiChoiceItems，设置为多选选项对话框
- setItems，设置为选项对话框，不区分多选单选
- setPositiveButton(int textId,DialogInterface.OnClickListener listener)
- setPositiveButton(CharSequence text, DialogInterface.OnClickListener listener)
- setNegativeButton(CharSequence text, DialogInterface.OnClickListener listener)
- setNegativeButton(int textId, DialogInterface.OnClickListener listener)
- setNeutralButton(CharSequence text, DialogInterface.OnClickListener listener)
- setNeutralButton(int textId, DialogInterface.OnClickListener listener)

以上方法均是为对话框设置按钮，且 positive、negative 和 neutral 类型的按钮最多只能设置一个。

setCustomTitle(View customTitleView)，设置自定义的 Title 视图。

setView(View view)，设置对话框内容为自定义的视图。

4.3.3　知识点 3：进度条（ProgressBar）

前面所学到的所有控件，有许多是为了来完成和用户交互而设计的，但是也有一些是为了增加用户体验而设计的，它们用来为程序提示信息、显示程序运行的状态等，就像上一节讲到的登

录时的对话框一样。本节讲的则是另外一个增加用户体验的控件，叫做进度条（ProgressBar）。提到进度条大家肯定都熟悉，我们在安装文件的时候都会出现一个提示安装进度的条，那便是进度条。

进度条与上一节我们讲到的 ProgressDialog 是不同的，第一，它继承自 Android.app.ProgressDialog；第二，在应用的时候，必须要新建 ProgressDialog 对象，在运行时会弹出"对话框"作为提醒，此时用户原来的应用程序就会失去焦点，直到进程结束后才会将控制权交给应用程序；第三，应用目标不同。在程序中我们可以将 ProgressBar 直接放到布局文件中并设置其属性为"不可见"，后来通过程序来将其设置为"可见"，并能获取程序运行的进度。

Android 系统提供了两个类进度条样式：长形进度条（ProgressBarStyleHorizontal）和圆形进度条（ProgressBarStyleLarge）。它们只是代表应用程序中某一部分程序的执行情况，而整个应用程序执行的进度，则可以通过应用程序标题栏来显示一个进度条，这就需要先对窗口的显示风格进行设置："requestWindowFeature(Window.FEATURE_PROGRESS);"。

下面我们就通过实例来学习进度条的用法和 Handler 的使用。我们在这个例子中将 ProgressBar 与 Handler 进行整合，通过 Handler 以及 Message 对象将进程里的状态往外传递，最后由 Activity 的 Handler 事件接收并取得运行状态。以下先来看一下实例运行的截图，如图 4-8 所示。

图 4-8 实例 CH4_3 运行图

要实现进度条，需要在布局文件中进行声明。其标签为<ProgressBar></ProgressBar>。布局文件源代码如代码清单 4-5 所示。

代码清单 4-5 第 4 章\CH4_3\res\layout\main.xml

```xml
<?xml version="1.0" encoding="utf-8"?>
<LinearLayout xmlns:android="http://schemas.android.com/apk/res/android"
    android:orientation="vertical"
    android:layout_width="fill_parent"
    android:layout_height="fill_parent"
    >
<TextView
    android:id="@+id/txtInfo"
    android:layout_width="fill_parent"
    android:layout_height="wrap_content"
    android:text="@string/hello"
    />
<ProgressBar
    android:id="@+id/progressBar1"
    style="?android:attr/progressBarStyleHorizontal"
    android:layout_width="200dp"
    android:layout_height="wrap_content"
    android:visibility="gone"/>
<ProgressBar
    android:id="@+id/progressBar2"
    style="?android:attr/progressBarStyleLarge"
    android:layout_width="wrap_content"
    android:layout_height="wrap_content"
    android:max="100"
    android:progress="50"
    android:secondaryProgress="70"
    android:visibility="gone"/>
<Button
    android:id="@+id/btnStart"
    android:layout_width="wrap_content"
    android:layout_height="wrap_content"
    android:text="start"/>
</LinearLayout>
```

在上面的代码中，style 表示进度条的风格，定义是长形进度条还是圆形进度条。Android:

max=""为设置进度条的最大值,android:progress=""为设置进度条的当前值,Android：visibility=""为设置进度条的可见性。在这里我们设置为"gone",在程序中通过按钮事件来改变其可见性的属性。

在程序中,为了能够让信息不断地传到 Activity 中,我们使用 Handler 对象和 Message 对象。为了能够控制进度条的进度,要声明两个整数：GUI_STOP_NOTIFIER 与 GUI_THREADING_NOTIFIER,作为信息传递出来时的信号标识。程序源代码如代码清单 4-6 所示。

代码清单 4-6 第 **4** 章**\CH4_3\src\sziit\practice\chapter4\CH4_3.java**
//包引入语句省略

```
public class CH4_3 extends Activity {//从Activity基类派生子类CH4_3
    // 声明变量
    private TextView txtInfo；//定义私有TextView对象
    private ProgressBar progressBar1；//定义私有ProgressBar对象
    private ProgressBar progressBar2；//定义私有ProgressBar对象
    private Button btnStart；//定义私有Button对象
    protected static final int GUI_STOP_NOTIFIER = 0x108；//定义静态只读int常量
    protected static final int GUI_THREADING_NOTIFIER = 0x109；//定义静态只读int常量
    public int intCounter = 0；//定义公有int变量
    public void onCreate(Bundle savedInstanceState) {//子类重写基类的onCreate方法
        super.onCreate(savedInstanceState)；//调用基类的onCreate方法
        requestWindowFeature(Window.FEATURE_PROGRESS)；// 设置窗口模式
        setProgressBarVisibility(true)；//显示进度条
        setContentView(R.layout.main)；//设置屏幕布局,通过R.layout.main引用布局资源
        txtInfo=(TextView)findViewById(R.id.txtInfo)；// 获得Progress对象
        progressBar1 = (ProgressBar) findViewById(R.id.progressBar1)；//查找ProgressBar对象
        progressBar2 = (ProgressBar) findViewById(R.id.progressBar2)；//查找ProgressBar对象
        btnStart = (Button) findViewById(R.id.btnStart)；//查找Button对象
        progressBar1.setIndeterminate(false)；//setIndeterminate 设置进度条是否自动运转
        progressBar2.setIndeterminate(false)；//禁止进度条自动运行
        // 设置按钮的事件监听器
        btnStart.setOnClickListener(new Button.OnClickListener() {
            public void onClick(View v) {//处理事件回调方法onClick
                progressBar1.setVisibility(View.VISIBLE)；//设置进度条为可见状态
                progressBar2.setVisibility(View.VISIBLE)；//设置进度条为可见状态
                progressBar1.setMax(100)；// 设置progressBar1的最大值
                progressBar1.setProgress(0)；//设置progressBar1的当前值
                progressBar2.setProgress(0)；//设置progressBar2的当前值
                // 通过线程来改变ProgressBar的值
```

```
                new Thread(new Runnable() {//创建进程
                    public void run() {//重写run方法
                        for (int i = 0; i < 10; i++) {
                            try {
                                intCounter = (i + 1) * 20;　//成员变量,用来识别加载进度
                                Thread.sleep(1000);　//循环一次,暂停一秒
                                //运行五秒后结束
                                if (i == 4) {//发送消息
                                    Message m = new Message();　//以Message对象,传递参数给Handler
                                    m.what = CH4_3.GUI_STOP_NOTIFIER;
                                    CH4_3.this.myMessageHandler.sendMessage(m);
                                    break;
                                } else {
                                    Message m = new Message();　//创建Message对象
                                    m.what = CH4_3.GUI_THREADING_NOTIFIER;
                                    CH4_3.this.myMessageHandler.sendMessage(m);
                                }   //发送消息
                            } catch (Exception e) {//捕获异常
                                e.printStackTrace();　//显示异常信息
                            }
                        }
                    }
                }).start();　//启动进程
            }
        });
    }
    Handler myMessageHandler = new Handler() {//创建Handler对象
        public void handleMessage(Message msg) {//重写handleMessage方法处理消息循环
            switch (msg.what) {//处理接收到的消息
                //到达最大值时
                case CH4_3.GUI_STOP_NOTIFIER:
                    txtInfo.setText(R.string.complete);
                    progressBar1.setVisibility(View.GONE);　//显示进度条
                    progressBar2.setVisibility(View.GONE);　//显示进度条
                    Thread.currentThread().interrupt();　//终止进程
                    break;
```

//处理过程中
case CH4_3.*GUI_THREADING_NOTIFIER*:
　　if (!Thread.*currentThread*().isInterrupted()) {
　　　　txtInfo.setText(
getResources().getText(R.string.*doing*)+"("+Integer.*toString*(intCounter)+"%)\n");
　　　　progressBar1.setProgress(intCounter);
　　　　progressBar2.setProgress(intCounter);
　　　　//设置标题栏中前景的一个进度条值
　　　　setProgress(intCounter * 100);
　　　　//设置标题栏中后面的一个进度条值
　　　　setSecondaryProgress(intCounter * 100);
　　}
　　break;
}
super.handleMessage(msg);　//调用基类handleMessage方法处理默认消息
　　}
};
}

在上面的程序中，progressBar1.setIndeterminate(**false**)为设置其是否自动运行。当为"ture"时进度条为不确定的进度即进度是未知的，对于长形的进度条则是一个动态的动画而不是进度。其运行截图如图 4-9 所示。

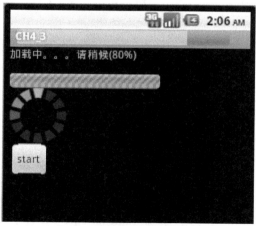

图 4-9　未知进度的进度条

对于圆形的进度条，因为其本身并不支持 indterminate mode 循环图片方式，所以在程序中要通过 TextView 将进度以文字的形式显示出来。另外，进度条的图片除了圆形之外，还有其他主题以及方形图片 Drawable 模式可以使用，感兴趣的读者可以自己学习。

4.3.4 知识点 4：菜单

Android 系统里面有 3 种类型的菜单：options menu、context menu、sub menu。

1．options menu

按 Menu 键就会显示 Option menu，用于当前的 Activity。它包括两种菜单项：icon menu 和 expanded menu。因为 options menu 在屏幕底部最多只能显示 6 个菜单项，这些菜单项称为 icon menu，它只支持文字（title）以及 icon，可以设置快捷键，不支持 checkbox 以及 radio 控件，所以不能设置 checkable 选项。而多于 6 个的菜单项会以"more" icon menu 来调出，称为 expanded menu，它不支持 icon，其他的特性都和 icon menu 一样。在 Activity 里面，一般通过以下方法来使用 options menu。

- Activity::onCreateOptionsMenu (Menu menu) 创建 options menu，这个方法只会在 menu 第一次显示时调用。
- Activity::onPrepareOptionsMenu (Menu menu) 更新改变 options menu 的内容，这个方法会在 menu 每次显示时调用。
- Activity::onOptionsItemSelected (MenuItem item) 处理选中的菜单项。

2．context menu

要在相应的 View 上按几秒后才显示的，用于 View，跟某个具体的 View 绑定在一起。这类型的菜单不支持 icon 和快捷键。在 Activity 里面，一般通过以下方法来使用 context menu。

- Activity::registerForContextMenu(View view) 为某个 View 注册 context menu，一般在 Activity::onCreate 里面调用。
- Activity::onCreateContextMenu(ContextMenu menu, View v, ContextMenu.ContextMenuInfo menuInfo) 创建 context menu，和 options menu 不同，context meun 每次显示时都会调用这个方法。
- Activity::onContextItemSelected(MenuItem item) 处理选中的菜单项。

3．sub menu

以上两种 menu 都可以加入子菜单，但子菜单不能嵌套子菜单，这意味着在 Android 系统，菜单只有两层，设计时需要注意，同时子菜单也不支持 icon。

4．xml 在菜单中的使用

上述的 3 种类型的 menu 都能够定义为 xml 资源，但需要手动地使用 MenuInflater 来得到 Menu 对象的引用。

一个菜单，对应一个 xml 文件，因为要求只能有一个根节点<menu>。xml 文件保存为 res/menu/some_file.xml。Java 代码引用资源: R.menu.some_file 接下来介绍相关的节点和属性(所有的属性都定义为 android 空间内，例如 android:icon="@drawable/icon")：

（1）<menu> 根节点，没有属性。

（2）<group> 表示在它里面的<item>在同一 group。相关属性包括：

① **id**：group id；

② **menuCategory**：对应常量 Menu CATEGORY_* ——定义了一组的优先权，有效值：

container、system、secondary 和 alternative；

③ **orderInCategory**：定义这组菜单在菜单中的默认次序，int 值；

④ **checkableBehavior**：定义这组菜单项是否 checkable，有效值：none、all(单选/单选按钮 radio button)、single(非单选/复选类型 checkboxes)；

⑤ **visible**：定义这组菜单是否可见，true or false；

⑥ **enabled**：定义这组菜单是否可用，true or false。

（3）<item> 菜单项，可以嵌入<menu>作为子菜单。相关属性包括：

① **id**：item id；

② **menuCategory**：用来定义 menu 类别；

③ **orderInCategory**：用来定义次序，与一个组在一起（Used to define the order of the item, within a group）；

④ **title**：标题；

⑤ **titleCondensed**：标题摘要，当原标题太长的时候，需要用简短的字符串来代替 title；

⑥ **icon**：icon 图标；

⑦ **alphabeticShortcut**：字母快捷键；

⑧ **numericShortcut**：数学快捷键；

⑨ **checkable**：是否为 checkbox，true or false；

⑩ **checked**：是否设置为 checked 状态，true or false；

⑪ **visible**：是否可见，true or false；

⑫ **enabled**：是否可用，true or false；

下面我们就来看一下一个 OptionMenu 和 ContextMenu 的简单范例程序，运行效果如图 4-10 和图 4-11 所示。

图 4-10　ContextMenu 范例

图 4-11　OptionMenu 范例

我们来分析下程序的代码，看要如何设置菜单。代码如代码清单 4-7 所示。

代码清单 4-7　第 4 章\CH4_4\ src\sziit\practice\chapter4\CH4_4.java

public class CH4_4 **extends** Activity {//从Activity基类派生子类CH4_4
　　public ImageView imageView；//声明ImageView的对象
　　public void onCreate(Bundle savedInstanceState) {//子类重写基类的onCreate方法
　　　　super.onCreate(savedInstanceState)；//调用基类的onCreate方法
　　　　setContentView(R.layout.*main*)；//设置屏幕布局，通过R.layout.*main*引用布局资源
　　　　imageView=(ImageView)findViewById(R.id.*imageView1*)；//为ImageView对象进行赋值
　　　　//为ImageView添加setOnCreateContextMenuListener监听，用以触发上下文菜单
　　　　imageView.setOnCreateContextMenuListener(**new**
ImageView.OnCreateContextMenuListener(){//设置OnCreateContextMenuListener监听器
　　　　　　public void onCreateContextMenu(ContextMenu menu, View arg1,
　　　　　　　　ContextMenuInfo arg2) {//处理回调方法onCreateContextMenu
　　　　　　　　menu.setHeaderIcon(R.drawable.*ic_action_search*);
　　　　　　　　menu.setHeaderTitle("图片");
　　　　　　　　menu.add(0, 0, 0, "搜索1");
　　　　　　　　menu.add(0, 0, 1, "搜索2");
　　　　　　　　menu.add(0, 0, 2, "搜索3");
　　　　　　}
　　　　});
　　}
　　//参考 android.app.Activity#onCreateOptionsMenu(android.view.Menu)

```java
//添加OptionMenu
public boolean onCreateOptionsMenu(Menu menu) {//重写onCreateOptionsMenu方法
    getMenuInflater().inflate(R.menu.optionsmenu, menu);
    return true;
}
public boolean onOptionsItemSelected(MenuItem item) {//重写onOptionsSelected方法
    super.onOptionsItemSelected(item); //调用基类onOptionsItemSelected方法
    switch(item.getItemId())//处理菜单项
    {
    case R.id.menu_1:
        Toast.makeText(CH4_4.this, "关于", Toast.LENGTH_SHORT).show();
        break;
    case R.id.menu_2:
        Toast.makeText(CH4_4.this, "选择", Toast.LENGTH_SHORT).show();
        break;
    case R.id.menu_3:
        finish();
        break;
    }
    return true;
}
```

在上面的代码中，我们通过代码为 ImageView 添加了上下文菜单，同时也添加了 OptionMenu，但是 OptionMenu 的菜单项并没有在代码中添加，而是借助于 xml 文件，在 xml 中进行了定义。如代码清单 4-8 所示。

代码清单 4-8 第 4 章\CH4_4\res\menu\optionmenu.xml

```xml
<menu xmlns:android="http://schemas.android.com/apk/res/android">
    <item android:id="@+id/menu_1"
        android:title="关于"
        android:orderInCategory="100" />
    <item android:id="@+id/menu_2"
        android:title="选择"
        android:orderInCategory="100" />
    <item android:id="@+id/menu_3"
        android:title="离开"
        android:orderInCategory="100" />
</menu>
```

4.4 项目实施

在本章中我们学习了 ListView 的使用。以下就来学习如何将所学的知识运用到项目中,如图 4-12 所示。

图 4-12 景点列表

在软件中我们使用了列表来将数据呈现给用户。但是又不仅仅是列表,而是在列表中放入了景点的略缩图,那么这又是如何实现的呢?另外,标题栏下面的各个按钮又是如何切换的呢?在数据加载时对话框又是如何显示的呢?下面就来分析其实现的方式。

首先分析下页面的布局,从标题栏开始,再就是类似 Tab 的按钮,然后以一个 ListView 来显示数据。那么它们的排列是用线性布局还是其他的呢?为了能够让程序适应不同分辨率的手机,我们建议使用相对布局,代码如代码清单 4-9 所示。

代码清单 4-9

```
<?xml version="1.0" encoding="utf-8"?>
<RelativeLayout xmlns:android="http://schemas.android.com/apk/res/android"
    android:id="@+id/root"
    android:background="@color/white"
    android:layout_width="fill_parent"
    android:layout_height="fill_parent" >
    <RelativeLayout
        android:id="@+id/titlebar_layout"
        android:layout_width="fill_parent"
        android:layout_height="wrap_content"
```

```
            android:background="@drawable/title_background"
            android:paddingLeft="5dip"
            android:paddingRight="5dip"
            android:paddingTop="5dip"
            android:paddingBottom="5dip"
            android:gravity="center"
            >
                <LinearLayout
                    android:orientation="horizontal"
                    android:layout_width="fill_parent"
                    android:layout_height="wrap_content"
                    android:layout_gravity="right|center_vertical"
                    >
                <TextView
                        android:textAppearance="?android:textAppearanceMedium"
                        android:layout_width="fill_parent"
                        android:layout_height="wrap_content"
                        android:text="@string/local_tour"
                        android:textSize="@dimen/title_textSize"
                        android:textColor="@color/white"
                        android:gravity="center"
                        />
                </LinearLayout>
        </RelativeLayout>
        <LinearLayout
                android:layout_below="@id/titlebar_layout"
                android:id="@+id/layoutBar"
                android:layout_width="fill_parent"
                android:layout_height="wrap_content"
                android:background="@drawable/big_button_up"
                android:orientation="horizontal" >
                <RelativeLayout
                    android:id="@+id/sense_layout"
                    android:layout_width="fill_parent"
                    android:layout_height="wrap_content"
                    android:layout_gravity="center_vertical"
                    android:layout_weight="1.0" >
                <TextView
```

```xml
            android:id="@+id/sense_tab"
            android:layout_width="fill_parent"
            android:gravity="center"
            android:layout_height="wrap_content"
            android:layout_centerInParent="true"
            android:textSize="@dimen/tab_textsize"
            android:textAppearance="?android:textAppearanceMedium"
            android:textColor="@color/black"
            android:text="@string/tour_scene" />
    </RelativeLayout>
    <RelativeLayout
        android:id="@+id/food_layout"
        android:layout_width="fill_parent"
        android:layout_height="wrap_content"
        android:layout_gravity="center_vertical"
        android:layout_weight="1.0" >
        <TextView
            android:id="@+id/food_tab"
            android:layout_width="fill_parent"
            android:gravity="center"
            android:layout_height="wrap_content"
            android:layout_centerInParent="true"
            android:textAppearance="?android:textAppearanceMedium"
            android:textColor="@color/black"
            android:textSize="@dimen/tab_textsize"
            android:text="@string/tour_food" />
    </RelativeLayout>
    <RelativeLayout
        android:id="@+id/quarter_layout"
        android:layout_width="fill_parent"
        android:layout_height="wrap_content"
        android:layout_gravity="center_vertical"
        android:layout_weight="1.0" >
        <TextView
            android:id="@+id/quarter_tab"
            android:layout_width="fill_parent"
            android:gravity="center"
            android:layout_height="wrap_content"
```

```xml
                android:layout_centerInParent="true"
                android:textAppearance="?android:textAppearanceMedium"
                android:textColor="@color/black"
                android:textSize="@dimen/tab_textsize"
                android:text="@string/tour_hotel" />
        </RelativeLayout>
        <RelativeLayout
            android:id="@+id/shopping_layout"
            android:layout_width="fill_parent"
            android:layout_height="wrap_content"
            android:layout_gravity="center_vertical"
            android:layout_weight="1.0" >
            <TextView
                android:id="@+id/shopping_tab"
                android:layout_width="fill_parent"
                android:gravity="center"
                android:layout_height="wrap_content"
                android:layout_centerInParent="true"
                android:textAppearance="?android:textAppearanceMedium"
                android:textColor="@color/black"
                android:textSize="@dimen/tab_textsize"
                android:text="@string/tour_shopping" />
        </RelativeLayout>
    </LinearLayout>
    <LinearLayout
        android:id="@+id/select_layout"
        android:layout_below="@id/layoutBar"
        android:orientation="horizontal"
        android:layout_height="30dip"
        android:layout_width="fill_parent"
        android:visibility="gone"
        android:gravity="center"
        >
        <LinearLayout
            android:layout_width="wrap_content"
            android:layout_height="wrap_content"
            android:layout_weight="1"
            android:gravity="center"
```

```xml
            >
            <Button
                android:id="@+id/type_btn"
                android:layout_width="wrap_content"
                android:layout_height="wrap_content"
                android:background="@drawable/btn_dropdown"
                android:textColor="@color/yellow"
                android:text="@string/tour_alltype"
                android:textSize="@dimen/sort_btn_textsize"
                />
        </LinearLayout>
        <LinearLayout
            android:layout_width="wrap_content"
            android:layout_height="wrap_content"
            android:layout_weight="1"
            android:gravity="center"
            >
            <Button
                android:id="@+id/grade_btn"
                android:layout_width="wrap_content"
                android:layout_height="wrap_content"
                android:background="@drawable/btn_dropdown"
                android:textColor="@color/yellow"
                android:text="@string/tour_allgrade"
                android:textSize="@dimen/sort_btn_textsize"
                />
        </LinearLayout>
        <LinearLayout
            android:layout_width="wrap_content"
            android:layout_height="wrap_content"
            android:layout_weight="1"
            android:gravity="center"
            >
            <Button
                android:id="@+id/id_sort"
                android:layout_width="wrap_content"
                android:layout_height="wrap_content"
                android:background="@drawable/btn_dropdown"
                android:textColor="@color/yellow"
```

```
                    android:text="@string/local_tour_sort"
                    android:textSize="@dimen/sort_btn_textsize"
                    />
        </LinearLayout>
    </LinearLayout>
    <LinearLayout
        android:id="@+id/loading_layout"
        android:orientation="horizontal"
        android:layout_width="fill_parent"
        android:layout_height="fill_parent"
        android:background="#ffffffff"
        android:gravity="center"
        android:padding="10dip"
        android:layout_below="@id/select_layout"
        >
        <ProgressBar
        android:id="@+id/progressBar1"
        android:layout_width="wrap_content"
        android:layout_height="wrap_content"
        />
        <View
                android:layout_width="10dip"
                android:layout_height="10dip"
                />
        <TextView
        android:layout_width="wrap_content"
        android:layout_height="wrap_content"
        android:textColor="#ff000000"
        android:text="载入中..."
        />
</LinearLayout>
        <include layout="@layout/network_erro"
            android:layout_below="@id/select_layout"
        />
        <ListView
            android:id="@+id/resultlist"
            android:layout_below="@id/select_layout"
            android:layout_height="match_parent"
```

```
        android:layout_width="match_parent"
        >

    </ListView>
</RelativeLayout>
```

4.5 技术拓展

上节我们学习了进度条的应用，这节将学习它的子类 SeekBar。SeekBar 类似于 ProgressBar，只是与 ProgressBar 不同的是，SeekBar 是用户可以控制的，如音乐播放器中的进度指示和调整工具、音量调节工具等。

由于 SeekBar 可以被用户控制，所以需要对其进行事件监听，因此需要实现 SeekBar.OnSeekBarChangeListener 接口。在 SeekBar 中需要设置 3 个事件监听，分别是：onProgressChanged（数值的改变）、onStartTrackingTouch（开始拖动）、onStopTrackingTouch（停止拖动），在 onProgressChanged 中我们可以通过设置得到当前的数值大小。下面就通过 SeekBar 来模仿调节音量的实例。首先还是来看一下程序运行的截图，如图 4-13 和图 4-14 所示。

图 4-13 SeekBar 调节中

图 4-14 SeekBar 调节完毕

在程序中使用 SeekBar，需要在布局文件中加入其标签，SeekBar 的标签为 <SeekBar></SeekBar>。布局文件的代码如代码清单 4-10 所示。

代码清单 4-10 第 4 章\CH4_5\res\layout\main.xml

```
<?xml version="1.0" encoding="utf-8"?>
<LinearLayout xmlns:android="http://schemas.android.com/apk/res/android"
    android:orientation="vertical"
    android:layout_width="fill_parent"
    android:layout_height="fill_parent"
    >
<SeekBar
    android:id="@+id/seekBar"
    android:layout_width="fill_parent"
    android:layout_height="wrap_content"
```

```
        android:max="100"
    />
    <TextView
        android:id="@+id/progress"
        android:layout_width="fill_parent"
        android:layout_height="wrap_content"
    />
<TextView
        android:id="@+id/tracking"
        android:layout_width="fill_parent"
        android:layout_height="wrap_content"/>
</LinearLayout>
```

在本程序中,由于没有设置SeekBar的当前值和secondaryProgress的值,我们可以通过SeekBar的属性:android:progress=""和 android:secondaryProgress=""来设置。我们通过 android:max="100"来将进度条的最大值设置为100。在使用时我们只要监听其事件并处理即可。

主程序的源代码如代码清单4-11所示。

代码清单4-11 第4章\CH4_5\src\sziit\practice\chapter4\CH4_5.java

```java
//从 Activity 基类派生子类 CH4_5,并实现 SeekBar.OnSeekBarChangeListener 接口
public class CH4_5 extends Activity implements SeekBar.OnSeekBarChangeListener {
    SeekBar seekBar;  //定义SeekBar对象
    TextView txtProgress;  //定义TextView对象
    TextView txtTracking;  //定义TextView对象
    public void onCreate(Bundle savedInstanceState) {//子类重写基类onCreate方法
        super.onCreate(savedInstanceState);  //调用基类onCreate方法
        setContentView(R.layout.main);  //设置屏幕布局,通过R.layout.main引用布局资源
        seekBar=(SeekBar)findViewById(R.id.seekBar);  //从布局中查找SeekBar对象
        seekBar.setOnSeekBarChangeListener(this);  //添加事件监听器
        txtProgress=(TextView)findViewById(R.id.progress);  //从布局中查找TextView对象
        txtTracking=(TextView)findViewById(R.id.tracking);  //从布局中查找TextView对象
    }
    //拖动中
    public void onProgressChanged(SeekBar seekBar, int progress,
            boolean fromUser) {//处理回调方法onProgressChanged
        txtProgress.setText("当前值:"+progress);  //显示当前执行进度
    }
    public void onStartTrackingTouch(SeekBar seekBar) {//处理回调方法onStartTrackingTouch
        txtTracking.setText("正在调节");
```

}
//停止拖动
public void onStopTrackingTouch(SeekBar seekBar) {//处理回调方法onStopTrackingTouch
　　txtTracking.setText("停止调节");
}
}

在程序中我们需要注意对 SeekBar 进行设置事件监听器，程序代码如下所示：

seekBar.setOnSeekBarChangeListener(**this**);

程序中，使用 onSartTrackingTouch()和 onStopTrackingTouch()来设置 TextView 中文字信息。

4.6 本章小结

在本章中我们介绍了 Android 的图形用户界面中比较高级的控件，包括 ListView、进度条、对话框以及菜单的使用。高级控件与基础控件相结合，合理的布局方式，是一个好的 Android 应用程序所必需的内容。

更多的图形用户界面需要读者自己去探索和学习，我们也可以自己定义控件，丰富应用程序。

4.7 强化练习

一、填空与简答题

1．在程序中可以通过_____方法来设置我们布局的背景颜色。
2．当轨迹球在 ListView 上滚动时，会调用 ListView 的_____方法。当单击 ListView 中的一项时会调用 ListView 的_____方法。
3．Android 平台中 ListView 显示的内容通过_____来与 ListView 实现对接。
4．请列举3种不同的对话框_____、_____、_____。
5．简述 Android 系统中的菜单类型，并说明它们之间的关系。

二、编程题

学习了本章的知识后，请大家实现如下的功能：为某一景点进行评分，选择星级指数，然后书写评语，最后通过单击"发表评论"按钮将数据放到我们的数据库中。

图 4-15 评价界面

第5章 Android 数据存储

5.1 项目导引

数据存储是应用程序最基本的问题，我们在日常的软件开发中不仅是企业系统，还是应用软件都需要解决这一问题。当然，作为一款手机应用，同样需要数据存储来保存用户的基本数据以及应用程序所需要的数据。如何灵活地使用数据存储机制，方便、快捷地让用户检索、查询和使用数据是我们需要考虑的。Android 一共为我们提供了 5 种不同的数据存储方式，分别如下：

（1）使用 SharedPreferences 存储数据；
（2）文件存储数据；
（3）SQLite 数据库存储数据；
（4）使用 ContentProvider 存储数据；
（5）网络存储数据。

SharedPreferences 主要是针对系统配置信息的保存，如给程序界面设置了音效，想在下一次启动时还能够保留上次设置的音效。由于 Android 系统的界面是采用 Activity 栈的形式，当系统内存不足时会回收一些 Activity 界面，所以有些操作需要在不活动时被保留下来，以便等再次激活时能够显示出来。又如，登录用户的用户名与密码，其采用了 Map 数据结构来存储数据，以键值的方式存储，可以简单地读取和写入。

文件存储数据顾名思义就是将要保存的数据以文件的形式保存。当需要调用所保存的数据时，只要读取这些文件就可以。

Android 内嵌了功能比其他手机操作系统强大的关系型数据库 sqlite3，我们学的 SQL 语句基本都可以使用，所创建的数据可以用 adb shell 来操作。具体路径是/data/data/package_name/databases。SQLite 是一个开源的关系型数据库，与普通关系型数据库一样，也具有 ACID 的特性。

Android 提供了 ContentProvider，一个程序可以通过实现一个 ContentProvider 的抽象接口将自己的数据完全暴露出去，而且 ContentProviders 是以类似数据库中表的方式将数据暴露的，也就是说，ContentProvider 就像一个"数据库"。那么外界获取其提供的数据，也就应该与从数据库中获取数据的操作基本一样，只不过它是采用 URI 来表示外界需要访问的"数据库"。

Android 中的网络存储数据就是将数据通过网络保存在网络上。在实际的使用中会使用 java.net.*和 android.net.*等类。

5.2 项目分析

在我们的项目中经常会遇到不同的情况，使用不同的存储方式。比如，保存用户的用户名和密码时可以使用 SharedPreferences 来保存，下次用户登录时不用再次输入用户名和密码即可登录。另外，随着 3G 网络的不断普及，手机客户端仅仅作为内容呈现的工具以及用户交互的工具，数据更多地则是存储在服务器并借助更大型的数据库来管理。在实际的项目开发中，我们需要根据实际的情况灵活选择不同的存储方式。下面就来学习一下 Android 下的存储方式。

5.3 技术准备

5.3.1 知识点1：文件存储

在 Android 中，可以在设备本身的存储设备或者外接的存储设备中创建用于保存数据的文件。在默认的情况下，文件是不能在不同的程序间共享的。在 5.1 节我们也提到了，由于 Android 是以 Linux 为内核的，所以其 File 的形式也是 Linux 下的形式。用文件来存储数据可以通过 openFileOutput 方法打开一个文件，然后通过 Load 方法获取文件中的数据，通过 deleteFile 方法删除一个指定的文件。

在进入 Files 的实例演示之前，我们需要了解什么是 Properties。Properties（属性），我们可以把 Properties 继承自 Hashtable，理解成一个 Hashtable，不过唯一不同的是，Properties 对应的"键—值"必须是字符串形式的数据类型。Files 数据存储主要是使用 Properties 配合 FileInputStream 或者 FileOutputStream 对文件进行写入操作。表 5-1 中列举出了一些 Properties 常用的方法。

表 5-1　　　　　　　　　　　　　　Properties 常用方法

方　　法	描　　述
getProperty(String name, String default-Value)	通过指定的 "name" 即 Key，搜索属性，第 2 个参数为默认值，即通过 Key 找不到文件中的属性时，要返回的默认值。返回值为 String
list(PrintStream out)	通过 PrintStream 列出可读的属性列表，无返回值

续表

方　　法	描　　述
list(PrintWriter)	通过 PrintWriter 列出可写的属性列表，无返回值
save(OutputStream out, String comment)	这种方法忽略任何 IO 异常，所以在实际操作过程中，可能会发生不必要的异常，故这种保存方法已过时，Google 并不推荐
setProperty(String name, String value)	设置属性，保存一个"键—值"对的属性
store(OutputStream out, String comment)	通过 FileOutputStream 打开对应的程序文件，然后通过 Store 保存之前 Properties 打包好的数据。这里备注可以为空
storeToXML(OutputStream os, String comment)	通过 FileOutputStream 打开对应的程序文件，将打包好的数据写入 XML 文件
storeToXML(OutputStream os, String comment, String encoding)	通过 FileOutputStream 打开对应的程序文件，将打包好的数据写入到 XML 文件，第 3 个参数可以指定编码

下面我们就来看一下 Files 在 Android 数据存储中的应用。实例 5-2 演示的是与实例 5-1 相同的功能，虽在布局上做了修改，但是所实现的功能是相同的。我们可以对比一下 SharedPreferences 和 Files 在实现上的区别。首先看一下效果图，如图 5-1 和图 5-2 所示。

图 5-1　程序启动初始界面

图 5-2　程序再次启动时默认账号和密码

下面就来看一下布局文件的代码，如代码清单 5-1 所示。

代码清单 5-1   第 5 章\CH5_1\res\layout\main.xml

```xml
<?xml version="1.0" encoding="utf-8"?>
<LinearLayout xmlns:android="http://schemas.android.com/apk/res/android"
 android:layout_width="fill_parent" android:layout_height="fill_parent"
 android:gravity="right" android:layout_gravity="right"
 android:background="@drawable/bg" android:orientation="vertical">
 <TableLayout android:layout_width="fill_parent"
  android:layout_height="wrap_content" android:stretchColumns="1">
  <TableRow android:gravity="center" android:layout_gravity="center">
   <ImageView android:layout_width="fill_parent"
    android:layout_height="wrap_content" android:id="@+id/ivlogo"
    >
   </ImageView>
  </TableRow>
 </TableLayout>
 <TableLayout android:layout_width="fill_parent"
  android:layout_height="wrap_content" android:stretchColumns="1">
  <TableRow android:layout_marginTop="100dip">
   <TextView android:layout_width="wrap_content"
    android:layout_marginLeft="20dip" android:gravity="center_vertical"
    android:layout_height="wrap_content" android:id="@+id/tvaccount"
    android:text="账号：" android:textSize="20sp">
   </TextView>
   <EditText android:layout_width="70px" android:layout_height="wrap_content"
    android:id="@+id/etaccount" android:layout_marginRight="20dip"
    android:maxLength="20"></EditText>
  </TableRow>
  <TableRow android:layout_marginTop="10dip">
   <TextView android:layout_width="wrap_content"
    android:layout_height="wrap_content" android:id="@+id/tvpw"
    android:layout_marginLeft="20dip" android:gravity="center_vertical"
    android:text="密码：" android:textSize="20sp">
   </TextView>
   <EditText android:layout_width="70px" android:layout_height="wrap_content"
    android:layout_marginRight="20dip" android:id="@+id/etpw"
    android:inputType="textPassword"></EditText>
```

```xml
    </TableRow>
   </TableLayout>
   <LinearLayout xmlns:android="http://schemas.android.com/apk/res/android"
      android:layout_width="wrap_content" android:layout_height="wrap_content"
      android:orientation="horizontal" android:layout_marginTop="5dip"
android:layout_marginRight="20dip">
      <TextView android:layout_width="wrap_content"
        android:layout_height="wrap_content" android:id="@+id/tvclear"
        android:text="清除Cookies" android:textColor="#aa0000" android:textSize="12px"></TextView>
   </LinearLayout>
   <TableLayout android:layout_width="fill_parent"
      android:layout_height="wrap_content" android:layout_marginTop="20dip">
      <TableRow android:gravity="center" android:layout_width="fill_parent">
        <Button android:layout_width="100px" android:layout_height="wrap_content"
          android:id="@+id/btnlogin" android:layout_gravity="center"
          android:text="登录"></Button>
        <Button android:layout_width="100px" android:layout_height="wrap_content"
          android:id="@+id/btnexit" android:layout_gravity="center"
          android:text="退出"></Button>
      </TableRow>
   </TableLayout>
   <LinearLayout xmlns:android="http://schemas.android.com/apk/res/android"
      android:layout_width="wrap_content" android:layout_height="wrap_content"
      android:orientation="horizontal" android:layout_marginTop="25dip">
      <CheckBox android:layout_width="wrap_content"
        android:layout_height="wrap_content" android:id="@+id/remeberPWD"
        android:text="记住密码" android:textSize="12px"></CheckBox>
      <CheckBox android:layout_width="wrap_content"
        android:layout_height="wrap_content" android:id="@+id/autoLogin"
        android:text="自动登录" android:textSize="12px"></CheckBox>
   </LinearLayout>
</LinearLayout>
```

上面的布局文件是多个样式的整合,正如前面我们说过的,可以借助工具来进行布局。下面来了解一下主程序的代码。

代码清单 5-2 第 5 章\CH5_1\ sziit\practice\study\chapter5\CH5_1

```java
public class CH5_1 extends Activity {//从Activity基类派生子类CH5_1
    private CheckBox remeberPWD;   //定义私有CheckBox对象
```

```java
    private CheckBox autoLogin;  //定义私有CheckBox对象
    private EditText extUser;  //定义私有EditText对象
    private EditText extPassword;  //定义私有EditText对象
    public void onCreate(Bundle savedInstanceState) {//子类重写基类onCreate方法
        super.onCreate(savedInstanceState);  //调用基类onCreate方法
        setContentView(R.layout.main);  //设置屏幕布局,通过R.layout.main引用布局资源
        remeberPWD=(CheckBox)findViewById(R.id.remeberPWD);  //查找CheckBox对象
        autoLogin=(CheckBox)findViewById(R.id.autoLogin);  //查找CheckBox对象
        extUser=(EditText)findViewById(R.id.etaccount);  //查找EditText对象
        extPassword=(EditText)findViewById(R.id.etpw);  //查找EditText对象
        load();  //从文件中读取数据
        remeberPWD
            .setOnCheckedChangeListener(new CompoundButton.OnCheckedChangeListener() {
                public void onCheckedChanged(CompoundButton buttonView,
                    boolean isChecked) {//处理回调方法onCheckedChanged
                if(remeberPWD.isChecked()){
                    save();  //保持数据
                }}
            });
        autoLogin
            .setOnCheckedChangeListener(new CompoundButton.OnCheckedChangeListener() {
                public void onCheckedChanged(CompoundButton buttonView,
                    boolean isChecked) {//处理回调方法onCheckedChanged
                }
            });
    }
//初始化操作
    private void load() {
        Properties properties=new Properties();  //新建Properties对象
        try{
            FileInputStream stream=this.openFileInput("user.cfg");  //打开文件读取数据
            properties.load(stream);  //从文件读取速度给properties对象
        }catch(FileNotFoundException e){//捕获没有找到文件或不能写入异常
            return;
        }catch(IOException e){//捕获文件读写异常
            return;
        }
```

```
        //取得数据
        extUser.setText(properties.getProperty("account").toString());   //读取属性数据
        extPassword.setText(properties.getProperty("password").toString());   //读取属性数据
    }
    private boolean save(){
        Properties properties=new Properties();   //创建Properties对象
        //将数据打包成Properties
        properties.put("account",extUser.getText().toString());
        properties.put("password",extPassword.getText().toString());
        try{
            FileOutputStream stream=this.openFileOutput("user.cfg",
Context.MODE_WORLD_WRITEABLE);  //创建FileOutputStream对象，向文件写数据
            properties.store(stream,"");   //将properties数据写入文件
        }catch(FileNotFoundException e){//捕获没有找到文件异常
            return false;
        }catch(IOException e){ //捕获文件读写异常
            return false;
        }
        return true;
    }
}
```

若文件不存在或不能以可写的方式打开，则会抛出 FileNotFoundException 异常。若文件不存在，也可以在打开文件的 OutputStream 同时创建文件，即使用 Context 对象调用 FileOutputStream openFileOutput(String name,int mode)。参数 name 为目录下文件名为 name 的 OutputStream，mode 为 Context 类中定义的一个常量，表示用什么模式打开或创建文件，当需要用到多个模式打开时用"|"隔开。其中 mode 的值如下：

- MODE_APPEND，以在文件末尾写入数据这种模式打开文件；
- MODE_PRIVATE，以仅仅只有应用程序自己可读可写的模式创建文件；
- MODE_WORLD_READABLE，以其他应用程序可读的模式创建文件；
- MODE_WORLD_WRITEABLE，以其他应用程序对文件可写的模式创建文件。

在大多数情况下，我们可以通过 Context 的对象调用 openFileOutput()和 openFileInput()来直接打开或创建位于私有目录的文件。另外，Context 对象还可以通过调用 fileList()方法来获得私有文件目录下所有文件的文件名组成的字符串数组。

与 SharedPreferences 的存储位置类似，我们用 files 来存储的数据也是放在/data/data/<包名>/files/下面，如图 5-3 所示。

Android 应用程序设计

Name	Size	Date	Time	Permiss...	Info
com.android.providers.applications		2012-08-01	06:42	drwxr-x--x	
com.android.providers.contacts		2012-08-19	04:22	drwxr-x--x	
com.android.providers.downloads		2012-08-19	04:24	drwxr-x--x	
com.android.providers.drm		2012-08-01	06:42	drwxr-x--x	
com.android.providers.media		2012-08-19	04:24	drwxr-x--x	
com.android.providers.settings		2012-08-19	04:22	drwxr-x--x	
com.android.providers.subscribedfe		2012-08-01	06:42	drwxr-x--x	
com.android.providers.telephony		2012-08-19	04:24	drwxr-x--x	
com.android.providers.userdictiona		2012-08-19	04:21	drwxr-x--x	
com.android.quicksearchbox		2012-08-19	04:24	drwxr-x--x	
com.android.sdksetup		2012-08-19	04:21	drwxr-x--x	
com.android.server.vpn		2012-08-01	06:42	drwxr-x--x	
com.android.settings		2012-08-19	04:22	drwxr-x--x	
com.android.soundrecorder		2012-08-01	06:42	drwxr-x--x	
com.android.spare_parts		2012-08-01	06:42	drwxr-x--x	
com.android.speechrecorder		2012-08-01	06:42	drwxr-x--x	
com.android.term		2012-08-19	04:21	drwxr-x--x	
com.android.wallpaper.livepicker		2012-08-19	04:21	drwxr-x--x	
com.example.android.apis		2012-08-19	04:22	drwxr-x--x	
com.example.android.livecubes		2012-08-19	04:21	drwxr-x--x	
com.example.android.softkeyboard		2012-08-19	04:21	drwxr-x--x	
com.study.chapter2		2012-12-21	02:23	drwxr-x--x	
com.study.chapter3		2012-12-21	03:02	drwxr-x--x	
com.study.chapter4		2012-12-24	01:38	drwxr-x--x	
com.study.chapter5		2012-12-24	04:38	drwxr-x--x	
files		2012-12-24	04:39	drwxrwx--x	
user.cfg	66	2012-12-24	04:39	-rw-rw---w-	
lib		2012-12-24	04:37	drwxr-xr-x	
com.svox.pico		2012-08-19	04:21	drwxr-x--x	
jp.co.omronsoft.openwnn		2012-12-24	04:39	drwxr-x--x	
dontpanic		2012-08-01	06:38	drwxr-x---	
local		2012-08-01	06:38	drwxrwx--x	
lost+found		2012-08-01	06:38	drwxrwx---	
misc		2012-08-01	06:38	drwxrwx--t	

图 5-3　files 存放位置

5.3.2　知识点 2：SharedPreferences

　　Shared Preferences 类似于我们常用的 ini 文件，用来保存我们在应用程序中的一些属性设置，在 Android 平台常用于存储简单的参数设置。就如上一节所讲到的，我们可以用来保存上一次用户所做的修改或者自定义参数设定，当再次启动程序时仍然保持原有的设置。下面就通过一个实例来看一下 SharedPreferences 的具体使用方法。在这个实例中，我们定义了两个 TextView 和两个 EditText，让用户输入用户名和密码，然后通过 Shared Preferences 使下次启动程序时自动默认的是用户设置过的用户名和密码。实例初始运行状态如图 5-4 所示，然后在文本框中分别输入 user 和 password，按返回键后，再次进入程序时界面如图 5-5 所示。

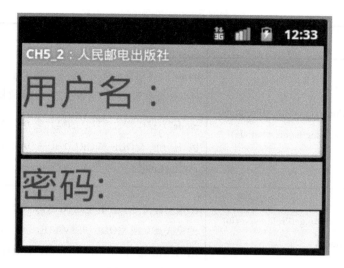

图 5-4　程序初始运行截图

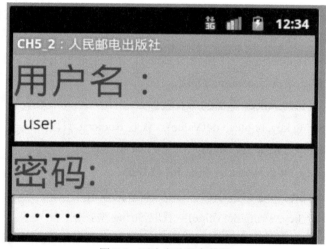

图 5-5　再次启动程序截图

使用 SharedPreferences 保存 key-value 的步骤如下。

（1）使用 Activity 类的 getSharedPreferences 方法获得 SharedPreferences 对象，其中存储 key-value 的文件的名称由 getSharedPreferences 方法的第一个参数指定。

（2）使用 SharedPreferences 接口的 edit 获得 SharedPreferences.Editor 对象。

（3）通过 SharedPreferences.Editor 接口的 putXxx 方法保存 key-value 对。其中，Xxx 表示不同的数据类型。例如，字符串类型的 value 需要用 putString 方法。

（4）通过 SharedPreferences.Editor 接口的 commit 方法保存 key-value 对。commit 方法相当于数据库事务中的提交（commit）操作。

表 5-2 中列出了两个获取 SharedPreferences 的方法。

表 5-2　　　　　　　　　　　　获取 SharedPreferences 的方法

返回值	函数	备注
SharedPreferences	Context.getSharedPreferences(String name,int mode)	name 为本组件的配置文件名（如果想要与本应用程序的其他组件共享此配置文件，可以用这个名字来检索到这个配置文件） mode 为操作模式，默认的模式为 0 或 MODE_PRIVATE，还可以使用 MODE_WORLD_READABLE 和 MODE_WORLD_WRITEABLE
SharedPreferences	Activity.getPreferences(int mode)	配置文件仅可以被调用的 Activity 使用。mode 为操作模式，默认的模式为 0 或 MODE_PRIVATE，还可以使用 MODE_WORLD_READABLE 和 MODE_WORLD_WRITEABLE

除此之外，以下再介绍几个重要的方法。

- public abstract boolean contains (String key)：检查是否已存在该文件，其中 key 是 xml 的文件名。
- edit()：为 preferences 创建一个编辑器 Editor，通过创建的 Editor 可以修改 preferences 里面的数据，但必须执行 commit()方法。
- getAll()：返回 preferences 里面的所有数据。
- getBoolean(String key, boolean defValue)：获取 Boolean 型数据。
- getFloat(String key, float defValue)：获取 Float 型数据。
- getInt(String key, int defValue)：获取 Int 型数据。
- getLong(String key, long defValue)：获取 Long 型数据。
- getString(String key, String defValue)：获取 String 型数据。
- registerOnSharedPreferenceChangeListener(SharedPreferences.OnSharedPreferenceChangeListener listener)：注册一个当 preference 发生改变时被调用的回调方法。
- unregisterOnSharedPreferenceChangeListener(SharedPreferences.OnSharedPreferenceChangeListener listener)：删除当前回调方法。

在理解了 SharedPreferences 的基本知识后我们来看一下程序的实现方法。由于布局文件比较简单，我们在这里就不再列出布局文件的代码清单，详细源代码请看本书所附光盘第 5 章/CH5_2 代码。主程序源代码如代码清单 5-3 所示。

代码清单 5-3 第 5 章\CH5_2\src\sziit\practice\chapter5\CH5_2.java

　　public class CH5_2 extends Activity {//从Activity基类派生子类CH5_2
　　public String userName,passWord;　//定义两个公有字符串String变量
　　public EditText extName,PWD；　//定义两个公有EditText对象
　　public void onCreate(Bundle savedInstanceState) {//子类重写基类onCreate方法

```
        super.onCreate(savedInstanceState);  //调用基类方法
        setContentView(R.layout.main);  //设置屏幕布局,通过R.layout.main引用布局资源
        extName=(EditText)findViewById(R.id.extUser);  //从布局资源中寻找EditText对象
        PWD=(EditText)findViewById(R.id.extPWD);  //从布局资源中寻找EditText对象
        SharedPreferences  user=getPreferences(Activity.MODE_PRIVATE);
        userName=user.getString("user_info", "");  //从SharedPreferences对象读取数据
        passWord=user.getString("password", "");  //从SharedPreferences对象读取数据
        extName.setText(userName);  //设置EditText显示内容
        PWD.setText(passWord);  //设置EditText显示内容
    }
    //当按下返回键时我们将在文本框中输入的内容保存到Preferences中
    public boolean onKeyDown(int keyCode,KeyEvent event){//处理回调方法onKeyDown
        if(keyCode==KeyEvent.KEYCODE_BACK){
            SharedPreferences user=getPreferences(0);  //获得SharedPreferences对象
            SharedPreferences.Editor editor=user.edit();
            editor.putString("user_info", extName.getText().toString());
            editor.putString("password",PWD.getText().toString());
            editor.commit();  //提交数据
            this.finish();
            return true;
        }
        return super.onKeyDown(keyCode, event);  //调用基类onKeyDown方法处理其他消息
    }
}
```

也许接下来你想知道我们保存的数据到底存放到什么地方了,那么下面就来看一下数据的保存。在 Eclipse 下切换到 DDMS 视图,选择 File Explorer 标签,找到/data/data 目录中对应的项目文件夹下的 Shared_prefs 文件夹。我们所看到的 xml 文件就是保存数据的地方了,如图 5-6 所示。

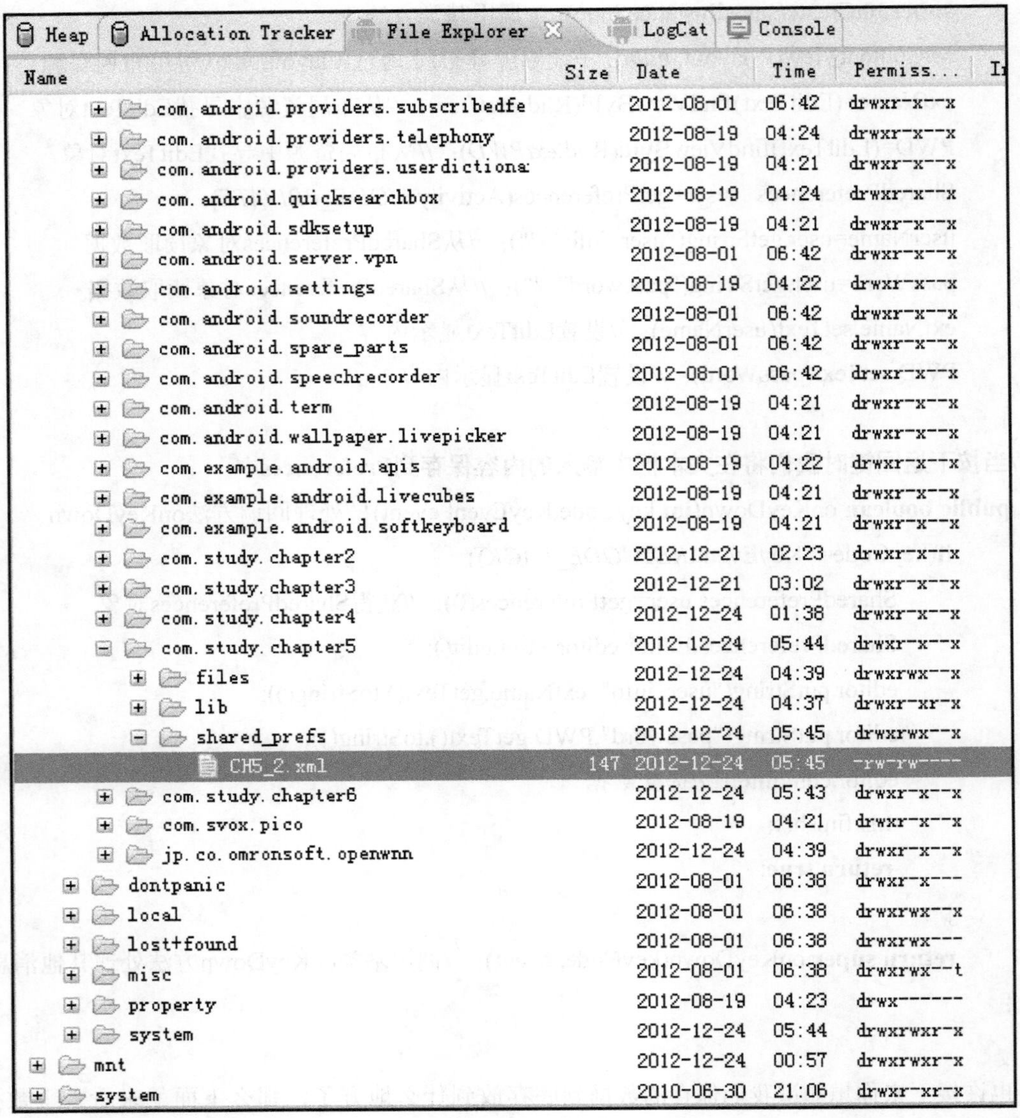

Android 应用程序设计

[File Explorer 截图]

图 5-6 Preferences 数据存储目录

5.3.3 知识点 3：嵌入式数据库 SQLite

前面我们已经讲解了 Android 平台下的两种数据存储的方式，但是这两种方式只是存储一些简单的、数据量较小的数据，当遇到大量的数据需要我们存储、管理的时候就需要用到经常使用的关系型数据库来存储数据。由于手机本身的局限性，我们不能使用在 PC 端所采用的 Oracle、SQL Server 等大型关系型数据库。Android 平台为开发者提供了 SQLite 数据库相关的 API 来实现对数据库的操作，用户可以很方便地使用这些 API 来对 SQLite 数据库进行创建、修改及查询等操作。本节我们将详细介绍 Android 平台下对 SQLite 的使用。

1. SQLite 数据库简介

SQLite 数据库是 D.Richard Hipp 用 C 语言编写的开源嵌入式数据库引擎。它的第一个 Alpha 版本诞生于 2000 年 5 月，是一款轻型数据库，主要针对嵌入式，占用的系统资源非常少但是却有强大的功能，据 Richard 保守估计 SQLite 可以处理每天 100 000 次单击率的 Web 站点，甚至可以处理上述数字的 10 倍负载。

SQLite 有许多的优点，首先是它的独立性，SQLite 数据库的核心引擎本身不依赖第三方软件，使用它不需要安装；第二，跨平台，SQLite 数据库支持大部分操作系统，除了在计算机上使用的操作系统之外，很多手机操作系统同样可以运行，如 Android、Windows Mobile、Symbian 等；第三，多语言接口，SQLite 数据库支持很多语言编程接口，如所使用的 C/C++、Java、Net 等。

上面只列出了 SQLite 几个显著特点，如果对 SQLite 的功能感兴趣可以浏览 SQLite 的官方网站:http://www.sqlite.org。

2. SQLite 操作详解

在 Android 中，我们通过 SQLiteDatabase 这个类的对象操作 SQLite 数据库。由于 SQLite 数据库并不需要建立连接以及身份验证，加上 SQLite 数据库单文件数据库的特性，使得获得 SQLiteDatabase 对象就像获得操作文件的对象那样简单。SQLite 数据库的一般操作包括：创建数据库、打开数据库、创建表、向表中添加数据、从表中删除数据、修改表中数据、关闭数据库、删除指定表、删除数据库和查询表中的某条数据。下面我们就一一来学习这些基本操作。

（1）创建和打开数据库

当我们要创建或者打开一个 SQLite 数据库时，可以直接调用 SQLiteDatabase 的静态方法：SQLiteDatabase openDatabase(String path，SQLiteDatabase.CursorFactory factory，int flags)，其中 path 是文件系统中数据的路径；CursorFactory 对象 factory 用于查询时构造 Cursor 的子类对象并返回，或者传入 null 使用默认的 factroy 构造；参数 flags 用于控制打开或者创建的模式，其中 flags 的值如下：

- OPEN_READONLY，只读的方式打开数据库；
- OPEN_READWRITE，可读写的方式打开；
- CREATE_IF_NECESSARY，数据库不存在时则创建数据库；
- NO_LOCALIZED_COLLATORS，打开数据库时，不根据本地化语言对数据进行排序。

当然，也可以调用方法 SQLiteDatabase openOrCreateDatabase(String path,SQLiteDatabase.CursorFactory factory)，此方法的作用与 openDatabase(path,factory,CREATE_IF_NECESSARY)相同。因为创建 SQLite 数据库也就是在文件系统中创建一个 SQLite 数据库的文件，所以应用程序必须对创建数据库的目录有可写的权限，否则会抛出 SQLiteException 异常。我们同样可以通过 Context 对象调用 SQLiteDatabase openOrCreateDatabase(String name,int mode,SQLiteDatabase.CursorFactroy factory)直接在私有数据库目录（/data/data/<package>/database/目录）中创建或打开一个名为 name 的数据库。注意：这里的 mode 并不是我们上面所说的 SQLiteDatabase 类中的几个整型常量，而是前面所提到的控制权限的常量：MODE_PRIVATE,MODE_WORLD_READABLE 和 MODE_WORLD_WRITEABLE。

除了在文件系统中创建 SQLite 数据库，Android 系统还支持 SQLite 内存数据库。在某些需要

143

临时创建数据库,并且对操作速率相对要求较高的情况下,SQLite 内存数据库就发挥作用了。如果要在内存中创建一个 SQLite 数据库,只需要调用 SQLiteDatabase 的静态方法 SQLiteDatabase create(SQLiteDatabase.CursorFactory factory),创建成功则返回创建的 SQLite 内存数据库的对象,否则返回 null。

注意:当我们获得的 SQLiteDatabase 对象不再使用时一定要调用 close()方法来关闭打开的数据库,否则会抛出 IllegalStateException 异常。

(2)通过 SQLiteDatabase 对象操作数据库

在 SQLite 数据库中除提供了 execSQL()和 rawQuery()这种直接对 SQL 语句解析的方法外,还针对 Insert、update、delete 和 select 等基本操作做了相关的定义,如表 5-3 所示。

表 5-3　　　　　　　　　　基本 SQLiteDatabase 操作语句

方法	参数	返回值	说明
public void execSQL(String sql); public void execSQL (String sql,Object[] bindArgs)	sql,需要执行的 sql 语句字符串; bindArgs,SQL 语句中表达式的?占位参数列表,仅仅支持 String、byte 数组、long 和 double 型数据作为参数	无	ExecSQL 能执行大部分的 sql 语句,在执行期间会获得该 SQLite 数据库的写锁,但是不支持用;隔开的多个 sql 语句,执行失败抛出 SQLException 异常
Public cursor rawQuery (String sql,String [] args); public Cursor rawQueryWithFactory (SQLiteDatabase.CursorFactory factory, String sql,String[] args,String editTable)	sql,需要执行的 sql 语句字符串; Args,?占位符只支持 String 类型; Factory,CursorFactory 对象,用来构造查询完毕时返回的 Cursor 的子类对象; editable,第一个可编辑的表名	Cursor 子类对象	执行一条语句并把查询结果以 Cursor 的子类对象的形式返回
Public long insert(String table,String nullColumnHack,ContentValues initialValues)	table,需要插入数据的表名; NullColumnHack,这个参数需要传入一个列名。SQL 标准并不允许插入所有列均为空一行数据,所以当传入的 initialValues 值为空或者为 0 时,用 nullColumnHack 参数指定的列会被插入值为 Null 的数据,然后再将此行插入到表中; InitalValues,用来描述要插入行数据的 ContentValues 对象,即列名和列值的映射	新插入行的 id	向表中插入一行数据

续表

方　　法	参　　数	返回值	说　　明
Public int update（String table，ContentValues values，String whereClause,String[] whereArgs）	table，需要更新数据的表名；Values，用来描述更新后的行数据的 ContentValues 对象，即列名和列值的映射；whereClause，用来指定所要更新的行；whereArgs,where 语句中表达式的？占位参数列表，只能为 String 类型	被更新的行的数量	更新表中指定行的数据
Public int delete(String talbe,String whereClause,String[] whereArgs)	table，需要删除数据的表名；whereClause,用来指定需要删除的行，若传入 null 则删除所有的行；whereArgs,where 语句中表达式的? 占位符，只能为 String 类型	若传入正确的 where 语句则返回被删除的行数，若传入 null,则返回 0	删除表中指定的行

5.4　项目实施

在上一章的学习中我们学习了如何布局我们的图形界面，那么如何来填充我们的数据呢？如图 5-7 所示的景点列表又是如何获取数据的呢？这就是本章所要解决的问题。详细代码参考本书项目实战源代码，这里主要来学习如何将数据信息填充到 ListView 之中。代码如下：

图 5-7　景点列表

```java
public void getData() {//获得数据
    LyaDao dao = null; //定义 LyaDao 对象
    dao = LyaUtil.getReadableDao(this);
    //List<LYW_JINGDIAN> sightData = dao.getLYW_JINGDIANList(null);
    sightlist = dao.getLYW_JINGDIANList(null);
    for (LYW_JINGDIAN sightdata : sightlist) {
        sightdata.loadData();
    }
    List<LYW_SHANGJIA> hotel = dao.getLYW_SHANGJIAList(null);
    hotelist = new ArrayList<LYW_SHANGJIA>();
    for (LYW_SHANGJIA hotelData : hotel) {
        hotelData.loadData();
        if (hotelData.getM_SHANGJIALEIXING_ID().equals("10")) {
            hotelist.add(hotelData);
        }
    }
    List<LYW_SHANGPIN> shangpin = dao.getLYW_SHANGPINList(null);
    foodlist = new ArrayList<LYW_SHANGPIN>();
    shoppinglist = new ArrayList<LYW_SHANGPIN>();
    for (LYW_SHANGPIN data : shangpin) {
        data.loadData();
        if (data.getM_SHANGPINDALEI_ID().equals("30")) {
            foodlist.add(data);
        } else if (data.getM_SHANGPINDALEI_ID().equals("20")) {
            shoppinglist.add(data);
        }
    }
    if (sightlist.size() == 0 || foodlist.size() == 0
            || hotelist.size() == 0 || shoppinglist.size() == 0) {
        loadingLinear.setVisibility(View.GONE);
        erroTxt.setVisibility(View.VISIBLE);
        erroTxt.setText("暂无数据");
    }
    switch (current) {
    case R.id.sense_tab:
        lourTourAdapter = new LourTourAdapter(this, sightlist);
        try {
```

```
                    lourTourAdapter.notifyDataSetChanged();
                    localTourListView.setAdapter(lourTourAdapter);
                } catch (Exception e) {//捕获异常
                    throw new RuntimeException("获取数据异常: " + e.getMessage());
                } finally {
                    try {
                        dao.close(); //关闭数据库
                    } catch (Exception e) {//捕获异常
                    }
                }
                break;
            case R.id.food_tab:
                lourTourFoodAdapter = new LourTourFoodAdapter(this, foodlist);
                try {
                    lourTourFoodAdapter.notifyDataSetChanged();
                    localTourListView.setAdapter(lourTourFoodAdapter);
                } catch (Exception e) {//捕获异常后,抛出运行时异常
                    throw new RuntimeException("获取数据异常: " + e.getMessage());
                } finally {
                    try {
                        dao.close(); //关闭数据库
                    } catch (Exception e) {//捕获异常
                    }
                }
                break;
            case R.id.quarter_tab:
                lourTourHotelAdapter = new LourTourHotelAdapter(this, hotelist);
                try {
                    lourTourHotelAdapter.notifyDataSetChanged();
                    localTourListView.setAdapter(lourTourHotelAdapter);
                } catch (Exception e) {//捕获异常后,抛出运行时异常
                    throw new RuntimeException("获取数据异常: " + e.getMessage());
                } finally {
                    try {
                        dao.close(); //关闭数据库
                    } catch (Exception e) {//捕获异常
                    }
```

```
            }
            break;
        case R.id.shopping_tab:
            lourTourFoodAdapter = new LourTourFoodAdapter(this, shoppinglist);
            try {
                lourTourFoodAdapter.notifyDataSetChanged();
                localTourListView.setAdapter(lourTourFoodAdapter);
            } catch (Exception e) {
                throw new RuntimeException("获取数据异常：" + e.getMessage());
            } finally {
                try {
                    dao.close();  //关闭数据库
                } catch (Exception e) {//捕获异常
                }
            }
            break;
    }
}
```

5.5 技术拓展

1. 数据共享（Content Providers）

Content Providers 是所有应用程序之间数据存储和检索的一个桥梁，当数据需要在应用程序之间共享时，我们就可以利用 Content Providers 为数据定义一个 URI，然后当其他应用程序对数据进行查询或者修改时，只需要从当前上下文对象获得一个 Content Resolver 传入相应的 URI 就可以了。在 Android 中，Conent Providers 是一种特殊的存储数据的类型，它提供了一套标准的接口来获取、操作数据。Android 系统本身也提供了几种常用的 Content Providers，如音频、视频、图像、个人联系信息等。程序通过 ContentProvider 访问数据而不需要关心数据具体的存储及访问过程，这样既提高了数据的访问效率，同时也保护了数据。

在学习 ContentProviders 之前需要先了解下 Content Reslover。上面说到了在 Android 中使用 Content Provider 来将应用程序自己的数据共享给其他应用程序，那么究竟是如何实现数据的共享呢？

Android 提供了 ContentProvider，一个程序可以通过实现一个 Content provider 的抽象接口将自己的数据完全暴露出去，而且 Content providers 是以类似数据库中表的方式将数据暴露。Content providers 存储和检索数据，通过它可以让所有的应用程序访问到，这也是应用程序之间唯一共享数据的方法。要想使应用程序的数据公开化，可通过两种方法：创建一个属于你自己的 Content provider 或者将你的数据添加到一个已经存在的 Content provider 中，前提是有相同数据类型并且有写入 Content provider 的权限。

如何通过一套标准及统一的接口获取其他应用程序暴露的数据？Android 提供了 ContentResolver，外界的程序可以通过 ContentResolver 接口访问 ContentProvider 提供的数据。但是在学习 ContentReslover 之前，我们需要重新了解下 URI 的相关知识。下面我们就一块来重新学习 URI 的知识。

2．URI 与 URL

在 Android 中广泛应用 URI，而不是 URL。URL 标识资源的物理位置，相当于文件的路径；而 URI 则是标识资源的逻辑位置，并不提供资源的具体位置。例如，电话薄中的数据，如果用 URL 来标识的话，可能会是一个很复杂的文件结构，而且一旦文件的存储路径改变，URL 也必须得改动。但是若是 URI，则可以用诸如 content : //contract /people 这样容易记录的逻辑地址来标识，而且并不需要关心文件的具体位置，即使文件位置改动也不需要变化，当然这都是对于用户来说，而后台程序中 URI 到具体位置的映射还是需要程序员来改动的。

我们先看下面这个例子：

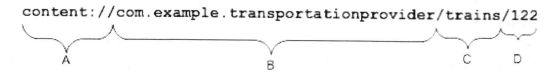

将其分为 A、B、C、D 4 个部分。

A：标准前缀，用来说明一个 Content Provider 控制这些数据，这部分是无法改变的。

B：URI 的标识，它定义了是哪个 Content Provider 提供这些数据。对于第三方应用程序，为了保证 URI 标识的唯一性，它必须是一个完整的、小写的类名。这个标识在<provider> 元素的 authorities 属性中说明，代码如下：

<provider name=".TransportationProvider" authorities="com.example.transportationprovider" ... >

C：路径，Content Provider 使用这些路径来确定当前需要产生什么类型的数据，URI 中可能不包括路径，也可能包括多个。

D：如果 URI 中包含此部分，表示需要获取的记录的 ID；如果没有 ID，就表示返回全部。

由于 URI 通常比较长，而且有时候容易出错，且难以理解。所以，在 Android 当中定义了一些辅助类，并且定义了一些常量来代替这些长字符串，如 People.CONTENT_URI。

3．Content Reslover

在掌握了 URI 的知识后我们来了解一下 ContentReslover。ContentResolver 是通过 URI 来查询 ContentProvider 中提供的数据。除了 URI 以外，还必须知道需要获取的数据段的名称，以及此数据段的数据类型。如果你需要获取一个特定的记录，就必须知道当前记录的 ID，也就是 URI 中 D 部分。应用程序通过一个唯一的 Content Reslover 接口来使用具体的某个 Content Providers。我们可以通过 getContentReslover()方法来取得一个 ContentReslover 对象，然后可以用 ContentReslover 提供的方法来操作 ContentProvider，如以下代码所示：

ContentReslover cr=getContentReslover();

查询初始化时，Android 系统会确定查询目标的那个 ContentProvider，并且确保它处于启动运行状态。系统会把所有 ContentProvider 对象实例化。通常，对于每一种类型的 ContentProvider 只有一个实例，但是这个实例可以与不同的程序或者进程中的多个 ContentReslover 对象进行通信。进程间的通信也是由 ContentReslover 和 ContentProvider 类处理的。

ContentResolver 也将采用类似数据库的操作来从 ContentProvider 中获取数据。现在简要介绍 ContentResolver 的主要接口，如表 5-4 所示。

表 5-4　　　　　　　　　　　ContentReslover 主要接口

返回值	函 数 声 明
final Uri	insert(Uri url, ContentValues values)Inserts a row into a table at the given URL.
final int	delete(Uri url, String where, String[] selectionArgs)Deletes row(s) specified by a content URI.
final Cursor	query(Uri uri, String[] projection, String selection, String[] selectionArgs, String sortOrder)Query the given URI, returning a Cursor over the result set.
final int	update(Uri uri, ContentValues values, String where, String[] selectionArgs)Update row(s) in a content URI.

Insert()方法用来插入数据，最后返回新插入数据的 URI。在此方法的实现中，只接受数据集的 URI，即指向表的 URI，然后利用数据库辅助对象获得的 SQLiteDatabase 对象，调用 insert()方法向指定表中插入数据。最后通知观察者数据已发生变化，返回插入数据的 URI。

Delete()和 update()方法用于数据的删除和修改操作，返回所影响数据的数目。

Query()方法是对数据进行查询的方法，最终将查询的结果包装入一个 Cursor 对象并返回。

4．ContentProvider

前面我们学习了如何使用 ContentProvider，现在我们来学习如何使用自己创建的 Content Provider。我们使用一个公共的静态常量 my_URI 来代表 URI 地址，需要注意，这个地址必须是唯一的。如以下代码段所示：

Public static final Uri my_URI=Uri.parse("content://com.android.study.myprovider");

我们需要定义返回给客户端的数据列名。根据使用的数据类型不同，我们的定义也会有所不同。若使用 Android 数据库，则数据列的使用方法和使用其他数据库一样，但是需要注意的是，我们需要定义一个名为_id 的列，用来作为主键，模式为 INTEGER PRIMARY KEY AUTOINCREMENT 自动更新。若我们使用的是一种新的数据类型，就要定义一个新的 MIME 类型，以供 ContentProvider.getType(url)来返回。MIME 类型也有两种不同形式：一个是为指定的单个记录的，还有一个为多条记录的。定义格式如下：

单个记录的 MIME 类型：

Vnd.android.cursor.item/vnd.companyname.contenttype

多个记录的 MIME 类型：

Vnd.android.cursor.dir/vnd.companyname.contenttype

我们在设置完后还需要在 AndroidManifest.xml 中使用<provider>标签来设置 ContentProvider。

设置方法如下代码段所示：

<provider android:name="MyContentProvider"
android:authorities=" sziit.practice.MyContentProvider"/>

当需要对其进行操作时同样需要在 AndroidManifest.xml 通过 android:readPermission()和 android:writePermission()来设置其操作权限，当然我们也可以使用 setReadPermission()和 setWritePermission()来设置。

5.6 本章小结

本章我们学习了 Android 系统中数据存储的不同方式，主要介绍了 SQLite 数据库的使用。在我们的开发中，恰当地选择数据存储方式将提高我们程序的用户体验。Android 系统中有 5 种不同的数据存储方式：文件存储、SharedPreferences、SQLite、网络存储以及 ContentProvider。其中，ContentProvider 主要用于程序间的数据存储与调用，网络存储主要用于访问后台数据以及将数据保存在网络服务中，是网络应用的首选。

5.7 强化练习

一、填空与简答题

1．Android 数据存储方式有_____种，分别是_____、_____、_____、_____和_____。

2．Android 中使用 SharedPreferences 使用_____的方式来存储数据。

3．简述使用 SharedPreferences 存储数据的基本步骤。

4．我们通过 openDatabase(String path,SQLiteDatabase.CursorFactory factory,int flags)方法打开数据库时，flags 的取值为_____、_____、_____和_____。

5．Android SQLite 数据库的操作方法有哪些？

二、编程题

编写 Android 应用程序，练习使用 SQLite 数据库。

第6章 Android 多媒体

6.1 项目导引

自从 Android 发布以来受到了人们的青睐，最重要的原因就是其开源性。我们可以在 Android 手机上装自己所喜欢的软件。而我们装的最多的无疑就是游戏、音乐播放、图片处理类的软件。这些都离不开 Android 系统的多媒体技术。本章我们就着重学习如何来将应用程序借助 Android 的多媒体技术做成一个"有声有色"的应用程序。

6.2 项目分析

在我们的实际开发中经常会用到多媒体的技术。在我们做的电子导游中需要语音播放我们所在景点的信息，界面如图 6-1 所示。

图 6-1　电子导游界面

在项目研发过程中我们需要考虑用户的使用状态。软件面向的是黄果树景区，在贵州，山地占大部分，故可能会存在网络信号差的问题，那么就需要考虑将我们的音频放到本地，直接调用。如果我们开发的是北京的电子导游，那么就可能需要将音频放到服务器上边下载边播放。

在这个项目中我们还碰到了一个问题，就是既然把音频放到了客户端，那么如何控制音频文件的大小就成了控制 apk 大小的决定因素。这也是我们在多媒体这章需要掌握和注意的。在实际操作之前先来学习下多媒体技术的基本内容。

6.3 技术准备

6.3.1 知识点 1：Android 网络基础（标准 Java、Apache、Android 网络和 HTTP 通信接口）

随着 3G 时代的来临，无论是上网、娱乐，还是办公、学习，智能手机将是用户的首选工具。Android 是一个以 Google 为首的由 30 多家科技公司和手机公司组成的开发手机联盟，受到广大手机用户的喜爱，由于丰富的应用程序作为支持，使得 3G 手机除了进行通话外，还给我们带来了更好的 3G 体验。

Android 目前有 3 种网络接口可以使用，分别为 java.net.*、org.apache 和 android.net.*。下面我们就简单地介绍一下这些接口的功能和作用。

1．标准 Java 接口

java.net.*提供与网络连接相关的类，java.net.*的包分为两个部分：低级 API 和高级 API。

（1）低级 API 主要用于处理以下抽象：

- 地址，也就是网络标识符，如 IP 地址；
- 套接字，也就是基本双向数据通信机制；
- 接口，用于描述网络接口。

（2）高级 API 主要用于处理以下抽象：

- URI，表示统一资源标识符；
- URL，表示统一资源定位符；
- 连接，表示到 URL 所指向资源的连接；
- 地址，在整个 java.net API 中，地址或者用作主机标识符或者用作套接字端点标识符。

下面我们通过一段程序代码来看下 java.net.*在程序中的使用，如代码清单 6-1 所示。

代码清单 6-1 使用 java.net 创建连接

```
try
{
    URL url=new URL("http://www.baidu.com");  //定义URL标识符
    HttpURLConnection http=(HttpURLConnection) url.openConnection();  //打开连接
    int nRC=http.getResponseCode();  //得到连接状态
    if(nRC==HttpURLConnection.HTTP_OK){
```

```
            //取得数据
            InputStream is=http.getInputStream();
            //处理数据
        }
}catch(Exception e){ //捕获异常
    }
```

2. Apache 接口

Http 协议是目前在 Internet 上使用最多、最重要的通信协议,越来越多的 Java 应用程序需要通过 Http 协议来访问网络资源。虽然前面我们讲到 java.net 包中已经提供了访问 Http 协议的基本功能,但是这对于大部分应用程序是不够的。Android 系统引入了 Apache HttpClient 以及对其的封装和扩展,如设置缺省的 Http 超时和缓存大小等。Android 使用的是目前最新的 HttpClient 4.0。通过 Apache 创建 HttpClient 以及 Get/Post、HttpRequest 等对象,设置连接参数,执行 Http 操作,处理服务器返回结果等功能。下面我们同样通过代码段来了解 Apache 接口的应用。

代码清单 6-2 使用 android.net.http.*连接网络

```
try{
    HttpClient hc=new DefaultHttpClient();   //创建HttpClient使用默认属性
    HttpGet get=new HttpGet("http://www.baidu.com");   //创建HttpGet实例
    HttpResponse rp=hc.execute(get);   //连接
    if(rp.getStatusLine().getStatusCode()==HttpStatus.SC_OK){
        InputStream is=rp.getEntity().getContent();
        ……//处理数据
    }
}catch(IOException ) //捕获异常
}
```

3. Android 网络接口

Android.net.*包实际上是通过对 Apache 的 HttpClient 进行封装,实现的一个 Http 变成接口,同时也提供了 Http 请求队列管理以及 Http 连接池管理,以提高并发情况下的处理效率,除此之外还有网络状态监视等接口、网络访问的 Socket、常用的 Uri 类以及有关 WiFi 相关的类等。代码清单 6-3 就是最简单的 Socket 连接代码,如下所示。

代码清单 6-3 Android 中的 Socket 链接

```
try{
    InetAddress inetAddress=InetAddress.getByName("192.168.1.25");   //获得InetAddress对象
    Socket client=new Socket(inetAddress,61203,true);   //创建套接字Socket对象
    InputStream in=client.getInputStream();   //获得Socket对象的输入流InputStream对象
    OutputStream out=client.getOutputStream();   //获得Socket对象的输出流OutputStream对象
    //处理数据
    out.close();   //关闭输出流OutputStream对象
```

in.close();　//关闭输入流InputStream对象
  client.close();　//关闭套接字Socket对象
}catch(UnknownHostException e){//捕获UnknownHostException异常
}catch(IOException e){//捕获IOException异常
}

4．HTTP 通信

  HTTP（Hyper Text Transfer Protocol，超文本传输协议）用于传送 WWW 方式的数据，采用了请求/响应模型。客户端向服务器发送一个请求，请求头包含了请求的方法、URI、协议版本以及包含请求修饰符、客户信息和内容的类似于 MIME 的消息结构。服务器以一个状态行作为响应，响应的内容包括消息协议的版本、成功或者错误编码，还包括服务器信息、实体元信息以及可能的实体内容。

  Google 以网络搜索引擎著称，自然而然也会使 Android SDK 拥有强大的 HTTP 访问能力。在 Android SDK 中，Google 集成了 Apache 的 HttpClient 模块。要注意的是，这里的 Apache HttpClient 模块是 HttpClient4.0（org.apache.http.*），而不是 Jakarta Commons HttpClient 3.x（org.apache.commons.httpclient.*）。

（1）Http Get 与 Http Post

  Http 通信中使用最多的就是 Get 和 Post。Get 请求方式中，参数直接放在 URL 字串后面，传递给服务器。

  格式如下：

  HttpGet method = new HttpGet("http://www.baidu.com?admin=Get");

  HttpResponse response = client.execute(method);

  而 Post 请求方式中，参数必须采用 NameValuePair[]数组的传送方式。

  格式如下：

  HttpPost method = new HttpPost("http://www.baidu.com");

  List<NameValuePair> params = new ArrayList<NameValuePair>();

  params.add(new BasicNameValuePair("admin", "Get"));

  method.setEntity(new UrlEncodedFormEntity(params));

  HttpResponse response = client.execute(method);

  在这两种通信方式中，一般情况下，两种方式实现的效果一样。但也有特殊情况，可能服务器只支持 GET 的请求方式，而不支持 POST 的请求方式，所以导致 POST 请求方式获取不到需要的数据；也可能服务器只支持 POST 的请求方式，不支持 GET 的请求方式。于是，我们需要查看服务器返回的状态码，如果是"200"则证明连接成功，否则连接失败。状态码的取得方式可以通过抓包观察，也可以直接用代码获取。

  用代码获得服务器返回的状态码具体参照为

  HttpResponse httpResponse = new DefaultHttpClient().execute(method);

  If(httpResponse.getStatusLine().getStatusCode() == 200)

{

// TODO: get data fran URL//从 URL 获取数据
}
else
{
　　// TODO: show connection false//显示连接异常信息
}

在这里我们需要注意的是，由于 Android 的很多操作都涉及权限的问题，如打电话和发短信等，都需要权限。而 Android 尝试连接网络时，也需要权限。加入网络连接权限：
在 AndroidManifest.xml 中添加
`<uses-permission android:name="android.permission.INTERNET" />`

（2）HttpURLConnection 接口

在 Android 的 SDK 中，Google 同时也继承了网络连接中标准的 Java 接口，使得最基本的一些连接方式得以继续沿用。注意，URLConnection 与 HttpURLConnection 都是抽象类，无法直接实例化对象。其对象主要通过 URL 的 openConnection 方法获得。

标准的 Java 接口格式如下：
URL url = new URL("http://www.baidu.com");　　//创建 URL 对象
HttpURLConnection http = (HttpURLConnection)url.openConnection();　　//获得对象实例
int response = http.getResponseCode();　　//获得服务器返回的状态码
if(200 == response)
{
　　//TODO: 从 URL 获得数据
}
else
{
　　// TODO: 显示连接异常信息
}

在上述这些方法中，如果所连接的网址不存在，会报出 java.net.UnknownHostException 异常，所以连接需要 try catch 捕获异常。

（3）网络接口

Android 中的网络接口，其实际上是通过对 Apache 中 HttpClient 的封装来实现的一个 HTTP 编程接口。

例子说明：

InetAddress ia = InetAddress .getByName("192.168.1.100");　　//获得 InetAddress 对象

Socket client = new Socket(ia, 8082, true);　　//创建套接字 Socket 对象

该例子已不适用于 Android，因为在 SDK 中，Socket 的构造方法 Socket(InetAddress addr, int port, boolean streaming) 被定义为： This constructor is deprecated。在 Android 中，被弃用的方法都调试不通，会报异常，所以该构造方法不能使用。

此处说明的是，因为在以前 Java 代码中（非 Android），有部分被定义为 This constructor is deprecated 的方法，被调用时并不会报异常，而使得部分程序员继续使用，在 Android 中是行不通的。

其实采用最简单的 Socket 的构造方法即可，例 Socket clientSocket =new Socket ("10.12.39.25", 9000)。

6.3.2　知识点 2：Service

Service 是在一段不定的时间运行在后台，不和用户交互的应用组件。每个 Service 必须在 manifest 中，通过<service>来声明，通过 contect.startservice 和 contect.bindserverice 来启动。

Service 和其他的应用组件一样，运行在进程的主线程中。这就是说，如果 Service 需要很多耗时或者阻塞的操作，则需要在其子线程中实现。

Service 的两种模式（startService()/bindService()不是完全分离的）：

- 本地服务 Local Service 用于应用程序内部。
  它可以启动并运行，直至有人停止了它或它自己停止。在这种方式下，它以调用 Context.startService() 启动，而以调用 Context.stopService() 结束；它可以调用 Service.stopSelf() 或 Service.stopSelfResult()来自己停止。不论调用了多少次 startService() 方法，只需要调用一次 stopService()来停止服务。用于实现应用程序自己的一些耗时任务，如查询升级信息，并不占用应用程序（如 Activity 所属线程），而是单开线程后台执行，这样用户体验比较好。

- 远程服务 Remote Service 用于 Android 系统内部的应用程序之间。
  它可以通过自己定义并暴露出来的接口进行程序操作。客户端建立一个到服务对象的连接，并通过那个连接来调用服务。连接以调用 Context.bindService()方法建立，以调用 Context.unbindService()关闭。多个客户端可以绑定至同一个服务。如果服务此时还没有加载，bindService()会先加载它。可被其他应用程序复用，比如天气预报服务，其他应用程序不需要再写这样的服务，调用已有的即可。

Service 的生命周期并不像 Activity 那么复杂，它只继承了 onCreate()、onStart()、onDestroy() 三个方法，当第一次启动 Service 时，先后调用了 onCreate()、onStart()这两个方法，当停止 Service 时，则执行 onDestroy()方法，这里需要注意的是，如果 Service 已经启动了，当我们再次启动 Service 时，不会再执行 onCreate()方法，而是直接执行 onStart()方法。启动 Service，根据 onStartCommand 的返回值不同，有两个附加的模式：

（1）START_STICKY 用于显示启动和停止 Service。

（2）START_NOT_STICKY 或 START_REDELIVER_INTENT 用于有命令需要处理时才运行的模式。

服务不能自己运行，需要通过调用Context.startService()或Context.bindService()方法启动服务。这两个方法都可以启动 Service，但是它们的使用场合有所不同。Context、Start Service()和 Context.bind Service()方法启动服务流程的对比如图 6-2 所示。

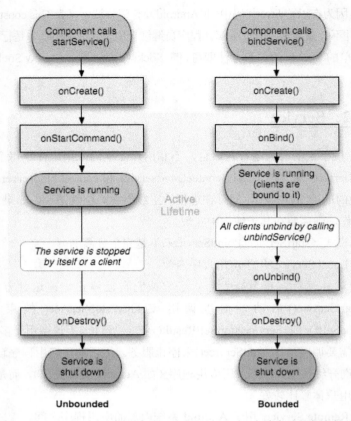

图 6-2　Context.startService()和 Context.bindService()方法启动服务流程对比

（1）使用 startService()方法启用服务，调用者与服务之间没有关联，即使调用者退出了，服务仍然运行。

如果打算采用 Context.startService()方法启动服务，在服务未被创建时，系统会先调用服务的 onCreate()方法，接着调用 onStart()方法。

如果调用 startService()方法前服务已经被创建，多次调用 startService()方法并不会导致多次创建服务，但会导致多次调用 onStart()方法。

采用 startService()方法启动的服务，只能调用 Context.stopService()方法结束服务，服务结束时会调用 onDestroy()方法。

（2）使用 bindService()方法启用服务，调用者与服务绑定在了一起，调用者一旦退出，服务也就终止，大有"不求同时生，必须同时死"的特点。

onBind()只有采用 Context.bindService()方法启动服务时才会回调。该方法在调用者与服务绑定时被调用，当调用者与服务已经绑定，多次调用 Context.bindService()方法并不会导致 on Bind()被多次调用。

采用 Context.bindService()方法启动服务时只能调用 onUnbind()方法解除调用者与服务解除，服务结束时会调用 onDestroy()方法。

一个 Service 可以同时 Start 并且 bind。如果 Service 已经 Start 或者 BIND_AUTO_CREATE 标

志被设置，则系统会一直保持 Service 的运行状态。如果没有一个条件满足，那么系统将会调用 onDestory 方法来终止 Service。所有的清理工作（终止线程，反注册接收器）都在 onDestory 中完成。

拥有 Service 的进程具有较高的优先级。只要 Service 已经被启动(start)或者客户端已连接(bindService)到它，则 Android 系统会尽量保持拥有 Service 的进程运行。当内存不足时，拥有 Service 的进程被保持的优先级较高。下面给出几种典型情形说明。

（1）如果 Service 正在调用 onCreate,onStartCommand 或者 onDestory 方法，那么用于当前 Service 的进程则变为前台进程以避免被杀掉（killed）。

（2）如果当前 Service 已经被启动(start)，拥有它的进程则比那些用户可见的进程优先级低一些，但是比那些不可见的进程更重要，这就意味着 Service 一般不会被 killed。

（3）如果客户端已经连接到 Service (bindService)，那么拥有 Service 的进程则拥有最高的优先级，可以认为 Service 是可见的。

（4）如果 Service 可以使用 startForeground(int, Notification)方法来将 Service 设置为前台状态，那么系统就认为是对用户可见的，并不会在内存不足时 killed。

另外，如果有其他的应用组件作为 Service，且 Activity 等运行在相同的进程中，那么将会增加拥有该组件的进程的重要性。

6.3.3 知识点 3：MediaPlayer

Android 提供了对常用音频和视频格式的支持。Android 支持的音频格式有 MP3、3GPP、Ogg 和 WAVE 等。支持的视频格式有 3GPP 和 MPEG 等。通过 Android 提供的 API 我们可以在应用中播放音频和视频，丰富应用。在本节中将着重介绍 MediaPlayer 在音频播放上的应用。

在 Android 中使用 MediaPlayer 类来播放音频比较简单，我们只需要创建一个 MediaPlayer 的对象，并指定要播放的音频文件，然后调用该类的 start()方法即可。

（1）创建 MediaPlayer 对象，并指定音频文件

通过 MediaPlayer 对象，可以使用 MediaPlayer 类提供的 create()方法来实现对音频文件的指定。其中 create()有两种使用方法，如下。

- create(Context context, int resID)

此方法用于从资源文件中加载静态的音频文件，并返回新建的 MediaPlayer 对象。

- create(Context context, Uri uri)

此方法用于根据我们指定的 URI 来动态装置音频文件，并返回新创建的 MediaPlayer 对象。

除了上文所讲的通过 create()方法搭载音频文件外，通过使用 MediaPlayer 的无参数的构造方法也可新建 MediaPlayer 对象，但在此种情况下，我们需要为 MediaPlayer 对象单独装载音频文件。步骤如下：

- 调用 MediaPlayer 对象的 setDataSource();

    如：MediaPlayer player=new MediaPlayer(); //创建 MediaPlayer 对象
        player.setDataSource("/sdcard/test.mp3"); //设置播放数据源

- 调用 MediaPlayer 对象的 prepare();

    如：player.prepare(); //做好播放准备

经过上述步骤后方才能将音频文件真正地加载到 MediaPlayer 对象中。

（2）开始播放

在上面的步骤完成以后，我们就可以使用调用 MediaPlayer 对象的 start()方法来开始播放或恢复播放处于暂停状态的音频，如 player.start()。

（3）停止播放

使用 MediaPlayer 对象提供的 stop()方法可以停止正在播放的音频，如 player.stop()。

（4）暂停播放

使用 MediaPlayer 类提供的 pause()方法可以暂停正在播放的音频。如：player.pause()。

MediaPlayer 生命周期

播放音频/视频文件和对流的控制是通过一个状态机来管理的。图 6-3 显示一个 MediaPlayer 对象被支持的播放控制操作驱动的生命周期和状态。椭圆代表 MediaPlayer 对象可能驻留的状态。弧线表示驱动 MediaPlayer 在各个状态之间迁移的播放控制操作。这里有两种类型的弧线。由一个箭头开始的弧代表同步的方法调用，而以双箭头开头的代表的弧线代表异步方法调用。

通过图 6-3，我们可以知道一个 MediaPlayer 对象有以下的状态。

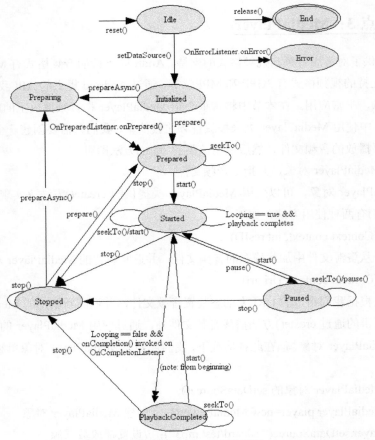

图 6-3　MediaPlayer 生命周期

（1）当一个 MediaPlayer 对象被 new 操作符创建或是调用了 reset()方法后，它就处于 Idle 状态。当调用了 release()方法后，它就处于 End 状态。这两种状态之间是 MediaPlayer 对象的生命周期。在一个新构建的 MediaPlayer 对象和一个调用了 reset()方法的 MediaPlayer 对象之间有一个微小的但是十分重要的差别。在处于 Idle 状态时，调用 getCurrentPosition()，getDuration()，getVideoHeight()，getVideoWidth()，setAudioStreamType(int)，setLooping(boolean)，setVolume(float, float)，pause()，start()，stop()，seekTo(int)，prepare() 或者 prepareAsync() 方法程序会出现错误。

当一个 MediaPlayer 对象刚被构建的时候，内部的播放引擎和对象的状态都没有改变，在这个时候调用以上的那些方法，框架将无法回调客户端程序注册的 OnErrorListener.onError()方法；但若这个 MediaPlayer 对象调用了 reset()方法之后，再调用以上的那些方法，内部的播放引擎就会回调客户端程序注册的 OnErrorListener.onError()方法，并将错误的状态传入。

一旦一个 MediaPlayer 对象不再被使用，应立即调用 release()方法来释放在内部的播放引擎中与这个 MediaPlayer 对象关联的资源。资源可能包括如硬件加速组件的单态组件，若没有调用 release()方法可能会导致之后的 MediaPlayer 对象实例无法使用这种单态硬件资源，从而退回到软件实现或运行失败。一旦 MediaPlayer 对象进入了 End 状态，它不能再被使用，也没有办法再迁移到其他状态。

此外，使用 new 操作符创建的 MediaPlayer 对象处于 Idle 状态，而那些通过重载的 create() 便利方法创建的 MediaPlayer 对象却不是处于 Idle 状态。事实上，如果成功调用了重载的 create() 方法，那么这些对象已经是 Prepare 状态了。

（2）一般情况下，由于种种原因一些播放控制操作可能会失败，如不支持音频/视频格式、缺少隔行扫描的音频/视频、分辨率太高、流超时等原因。因此，错误报告和恢复在这种情况下是非常重要的。有时，由于编程错误，在处于无效状态的情况下调用了一个播放控制操作可能发生。在所有这些错误条件下，内部的播放引擎会调用一个 OnErrorListener.onError()方法。一旦发生错误，MediaPlayer 对象会进入 Error 状态。为了重用一个处于 Error 状态的 MediaPlayer 对象，可以调用 reset()方法来把这个对象恢复成 Idle 状态。注册一个 OnErrorListener 来获知内部播放引擎发生的错误是一个好的编程习惯。在不合法的状态下调用一些方法，如 prepare()，prepareAsync() 和 setDataSource()方法会抛出 IllegalStateException 异常。

（3）调用 setDataSource(FileDescriptor)方法或 setDataSource(String)方法，或 setDataSource(Context，Uri)方法，或 setDataSource(FileDescriptor, long, long)方法会使处于 Idle 状态的对象迁移到 Initialized 状态。若当此 MediaPlayer 处于其他的状态下，调用 setDataSource()方法，会抛出 IllegalStateException 异常。好的编程习惯是不要疏忽了调用 setDataSource()方法的时候可能会抛出的 IllegalArgumentException 异常和 IOException 异常。

（4）在开始播放之前，MediaPlayer 对象必须要进入 Prepared 状态。有两种方法（同步和异步）可以使 MediaPlayer 对象进入 Prepared 状态：一种是调用 prepare()方法（同步），此方法返回就表示该 MediaPlayer 对象已经进入了 Prepared 状态；另一种是调用 prepareAsync()方法（异步），此方法会使此 MediaPlayer 对象进入 Preparing 状态并返回，而内部的播放引擎会继续未完成的准备工作。当同步版本返回时或异步版本的准备工作完全完成时就会调用客户端程序员提供的 OnPreparedListener.onPrepared()监听方法。可以调用 MediaPlayer.setOnPreparedListener(android.

media.MediaPlayer.OnPreparedListener)方法来注册 OnPreparedListener。Preparing 是一个中间状态，在此状态下调用任何方法的结果都是未知的！ 在不合适的状态下调用 prepare()和 prepareAsync()方法会抛出 IllegalStateException 异常。当 MediaPlayer 对象处于 Prepared 状态的时候，可以调整音频/视频的属性，如音量、播放时是否一直亮屏、循环播放等。

（5）要开始播放，必须调用 start()方法。当此方法成功返回时，MediaPlayer 的对象处于 Started 状态。isPlaying()方法可以被调用来测试某个 MediaPlayer 对象是否在 Started 状态。当处于 Started 状态时，内部播放引擎会调用客户端程序员提供的 OnBufferingUpdateListener。onBufferingUpdate()回调方法，此回调方法允许应用程序追踪流播放的缓冲状态。对一个已经处于 Started 状态的 MediaPlayer 对象调用 start()方法没有影响。

（6）播放可以被暂停、停止，以及调整当前播放位置。当调用 pause()方法并返回时，会使 MediaPlayer 对象进入 Paused 状态。注意 Started 与 Paused 状态的相互转换在内部的播放引擎中是异步的，所以可能需要一点时间在 isPlaying()方法中更新状态，若正在播放流内容，这段时间可能会有几秒钟。调用 start()方法会让一个处于 Paused 状态的 MediaPlayer 对象从之前暂停的地方恢复播放。当调用 start()方法返回的时候，MediaPlayer 对象的状态会又变成 Started 状态。对一个已经处于 Paused 状态的 MediaPlayer 对象，pause()方法没有影响。

（7）调用 stop()方法会停止播放，并且还会让一个处于 Started、Paused、Prepared 或 PlaybackCompleted 状态的 MediaPlayer 进入 Stopped 状态。对一个已经处于 Stopped 状态的 MediaPlayer 对象，stop()方法没有影响。

（8）调用 seekTo()方法可以调整播放的位置。seekTo(int)方法是异步执行的，所以它可以马上返回，但是实际的定位播放操作可能需要一段时间才能完成，尤其是播放流形式的音频/视频。当实际的定位播放操作完成之后，内部的播放引擎会调用客户端程序员提供的 OnSeekComplete.onSeekComplete()回调方法。可以通过 setOnSeekCompleteListener(OnSeek CompleteListener)方法注册。需要注意的是，seekTo(int)方法也可以在其他状态下调用，如 Prepared、Paused 和 PlaybackCompleted 状态。此外，目前的播放位置，实际可以调用 getCurrentPosition()方法得到，它可以帮助诸如音乐播放器的应用程序不断更新播放进度。

（9）当播放到流的末尾，播放就完成了。如果调用了 setLooping(boolean)方法开启了循环模式，那么这个 MediaPlayer 对象会重新进入 Started 状态；若没有开启循环模式，那么内部的播放引擎会调用客户端程序员提供的 OnCompletion.onCompletion()回调方法，可以通过调用 MediaPlayer.setOnCompletionListener(OnCompletionListener)方法来设置。内部的播放引擎一旦调用了 OnCompletion.onCompletion()回调方法，说明这个 MediaPlayer 对象进入了 PlaybackCompleted 状态。当处于 PlaybackCompleted 状态的时候，可以再调用 start()方法来让这个 MediaPlayer 对象再进入 Started 状态。

最后通过一个实例来看一下在 Android 中音频的播放。我们自己来制作一个简易的 MP3 播放器。源码为 CH6_1，效果图如图 6-4 所示。

图 6-4　Android 音乐播放器

在布局文件的设计中我们需要添加一个<ListView>，以供用来列出所有的 Sdcard 中的文件。布局文件的源代码如代码清单 6-4 所示。

代码清单 6-4　第 6 章\CH6_1\res\layout\main.xml

```xml
<?xml version="1.0" encoding="utf-8"?>
<AbsoluteLayout android:id="@+id/widget0"
    android:layout_width="fill_parent" android:layout_height="fill_parent"
    xmlns:android="http://schemas.android.com/apk/res/android"
    android:background="@drawable/background">
    <ImageButton android:id="@+id/pause"
        android:layout_width="32px" android:layout_height="32px"
        android:layout_x="141px" android:layout_y="44px"
        android:src="@drawable/pause">
    </ImageButton>
    <ImageButton android:id="@+id/last"
        android:layout_width="32px" android:layout_height="32px"
        android:layout_x="75px" android:layout_y="112px"
        android:src="@drawable/lastone">
    </ImageButton>
    <ImageButton android:id="@+id/next"
        android:layout_width="32px" android:layout_height="32px"
        android:layout_x="207px" android:layout_y="112px"
        android:src="@drawable/nextone">
```

```xml
</ImageButton>
<ImageButton android:id="@+id/start"
    android:layout_width="48px" android:layout_height="48px"
    android:layout_x="133px" android:layout_y="105px"
    android:src="@drawable/start">
</ImageButton>
<ImageButton android:id="@+id/stop"
    android:layout_width="32px" android:layout_height="32px"
    android:layout_x="141px" android:layout_y="180px"
    android:src="@drawable/stop">
</ImageButton>
<ListView
android:id="@id/android:list"
android:layout_width="wrap_content"
android:layout_height="wrap_content"
android:layout_weight="1"
android:drawSelectorOnTop="false"/>
</AbsoluteLayout>
```

最后来看一下在主程序中音乐播放器的代码清单 6-5。

代码清单 6-5 第 6 章\CH6_1\src\sziit\practice\chapter6\CH6_1.java

```java
public class CH6_1 extends ListActivity {//从ListActivity基类派生子类CH6_1
    private ImageButton btnLast,btnStart,btnNext,btnStop,btnPause; //定义私有ImageButton对象
    private boolean bIsReleased=false;
    private boolean bIsPaused=false;
    private MediaPlayer mPlayer=new MediaPlayer(); //创建私有MediaPlayer对象
    //音乐列表
    private List<String> myMusic=new ArrayList<String>(); //创建私有List<String>对象
    //当前播放歌曲的索引
    private int current=0;
    private static final String path=new String("/sdcard/"); //路径path
    public void onCreate(Bundle savedInstanceState) {//子类重写基类onCreate方法
        super.onCreate(savedInstanceState); //基类调用onCreate方法
        setContentView(R.layout.main); //设置屏幕布局，通过R.layout.main引用屏幕资源
        btnLast=(ImageButton)findViewById(R.id.last); //从布局中查找ImageButtond对象
        btnStart=(ImageButton)findViewById(R.id.start); //从布局中查找ImageButtond对象
        btnNext=(ImageButton)findViewById(R.id.next); //从布局中查找ImageButtond对象
        btnStop=(ImageButton)findViewById(R.id.stop); //从布局中查找ImageButtond对象
```

```
btnPause=(ImageButton)findViewById(R.id.pause);  //从布局中查找ImageButtond对象
MusicList();
//设置ImageButton对象的单击事件监听器OnClickListener：
btnStart.setOnClickListener(new ImageButton.OnClickListener(){
    public void onClick(View v) {//处理单击回调方法onClick
        playMusic(path+myMusic.get(current));  //播放音乐
    }
});
//暂停播放:设置btnPause的单击事件监听器OnClickListener
btnPause.setOnClickListener(new ImageButton.OnClickListener(){
    public void onClick(View v)
    {//处理单击回调方法onClick
        if(mPlayer!=null){
            if(bIsReleased==false){
                if(bIsPaused==false){
                    mPlayer.pause();  //停止播放
                    bIsPaused=true;
                }else if(bIsPaused==true){
                    mPlayer.start();  //开始播放
                    bIsPaused=false;
                }
            }
        }
    }
});
//next(下一首)：设置btnNext的单击事件监听器OnClickListener
btnNext.setOnClickListener(new ImageButton.OnClickListener(){
    public void onClick(View v) {//处理单击回调方法onClick
        nextMusic();}//播放最后一首音乐
});
//lastOne
btnLast.setOnClickListener(new ImageButton.OnClickListener(){
    public void onClick(View v)
        lastMusic();
    }
});
//stop（停止播放）：设置btnStop的单击事件监听器OnClickListener
```

```
            btnStop.setOnClickListener(new ImageButton.OnClickListener(){
                public void onClick(View v) {//处理单击回调方法onClick
                    if(mPlayer.isPlaying()==true){
                        mPlayer.reset();  //播放器复位
                    }
                }
            });
        }
        //绑定音乐
        void MusicList(){//将音乐文件添加到播放列表中
            File home=new File(path);  //创建文件File对象
        if(home.listFiles().length>0){//遍历目录中所有文件
            for(File file:home.listFiles()){
                myMusic.add(file.getName());  //添加到播放器列表
            }
            ArrayAdapter<String> musicList=new ArrayAdapter<String>
(CH6_1.this,R.layout.musicitme, myMusic);
            setListAdapter(musicList);
        }
        }
        //播放音乐
        void playMusic(String path){
            try {
            mPlayer.reset();  //播放器复位
            mPlayer.setDataSource(path);  //设置播放数据来源
            mPlayer.prepare();  //作播放准备
            mPlayer.start();  //开始播放
            mPlayer.setOnCompletionListener(new OnCompletionListener() {//设置监听器
                public void onCompletion(MediaPlayer mp) {
                }
            });
            } catch (Exception e) {//捕获异常
                e.printStackTrace();  //显示异常信息
            }
        }
        //下一首
        void nextMusic(){
```

166

```
        if(++current>=myMusic.size()){
            current=0;
        }
        else{
            playMusic(path+myMusic.get(current));
        }
    }
    //上一首
    void lastMusic(){
        if(current!=0)
            {
        if(--current>=0){
            current=myMusic.size();
        } else{
            playMusic(path+myMusic.get(current));
        }
            } else{
            playMusic(path+myMusic.get(current));
        }
    }
    //当用户返回时结束音乐并释放音乐对象
        public boolean onKeyDown(int keyCode, KeyEvent event) {//重写onKeyDown方法
            if(keyCode==KeyEvent.KEYCODE_BACK){//处理KEYCODE_BACK按键
                mPlayer.stop();   //停止播放
                mPlayer.release();   //释放播放器占用资源
                this.finish();
                return true;
            }
        return super.onKeyDown(keyCode, event);  //调用基类onKeyDown方法处理其他按键
    }
        //当选择列表项时播放音乐
        protected void onListItemClick(ListView l, View v, int position, long id) {
         current=position;
         playMusic(path+myMusic.get(current));
        }
}
```

在这个实例中，有两点需要注意，第一， Activity 是继承自 ListActivity；第二，需要在模拟

器的 Sdcard 中放入几个音乐文件以作测试。在向模拟器中的 SD 卡存储文件时有两种方法：第一种方法是通过 cmd，要 cd 到 android 的 SDK 的 Tools 目录下，用 adb push 命令添加如 E:\skyland\android-sdk-windows-1.0_r2\tools>adb push new.JPG /sdcard，第一个参数为要加入的图片（mp3）全名，如果名字中间有空格，要用双引号将其括起来，如 E:\skyland\android-sdk-windows-1.0_r2\tools>adb push "First Start.mp3" /sdcard，第二参数就是我们创建的 sdcard；第二种方法为首先启动程序，运行模拟器；在 Eclipse 上先打开 DDMS 窗口，选择 File Explorer 标签，选中右上角的两个箭头就变成可用，右边箭头是导入，左边箭头是导出。

6.3.4 知识点 4：视频

视频播放比音频播放要复杂一些，除了音频之外，还需要考虑视觉组件。为了解决这一问题，Android 提供了一个专门的视图控制器 android.widget.VideoView，封装了 MediaPlayer 的创建和初始化过程，这个小部件可以用在任何布局管理器中，而且它还提供了很多显示选项，包括缩放和着色。要实现播放功能，我们要做的是，创建一个 VideoView 小部件并将其设置为用户界面的内容，然后设置视图的路径或 URI 并触发 start()方法。视频播放除了需要按钮控件外，还需要一个显示视频的框架。对于我们这个例子，这里使用 VideoView 组件显示视频内容，没有创建我们自己的按钮控件，而是创建了一个 MediaController 来提供这些按钮（如果需要另外创建的话，也可以自己创建）。

下面通过一个实例来学习 Android 为我们提供的视频播放控件的使用。VideoView 的使用与之前所学过的 ImageView 的使用是类似的，区别在于 ImageView 用来存放的是图片，而 VideoView 则显示的是 Video。还有一点需要注意的是，在 VideoView 中需要对视频进行控制，所以要搭配 MediaController。首先来看一下实例的效果图，如图 6-5 所示。

图 6-5　VideoView 视频播放

我们通过在布局文件中加入视频控件，就能够在程序中使用视频播放控件了。实例的布局文

第 6 章 Android 多媒体

件如代码清单 6-6 所示。

代码清单 6-6 第 6 章\CH6_2\res\layout\main.xml

<?xml version=*"1.0"* encoding=*"utf-8"*?>
<LinearLayout xmlns:android=*"http://schemas.android.com/apk/res/android"*
    android:orientation=*"vertical"*
    android:layout_width=*"fill_parent"*
    android:layout_height=*"fill_parent"*
    >
<VideoView android:id=*"@+id/videoView"*
android:layout_width=*"fill_parent"*
android:layout_height=*"wrap_content"*
android:layout_centerInParent=*"true"*/>
</LinearLayout>

在程序中，我们还是首先要获取到 VideoView 的对象，然后新建 MediaController 来控制视频播放器。在使用 VideoView 时我们可以通过 setVideoPath(String path)方法为视频播放器加载视频文件，然后通过 setMediaController（MediaController controller）来让 VideoView 获得焦点。主程序源码如代码清单 6-7 所示。

代码清单 6-7 第 6 章\CH6_2\src\sziit\practice\chapter6\CH6_2.java
public class CH6_2 **extends** Activity {//从Activity基类派生子类CH6_2
　　public VideoView myVideoView；//定义公有VideoView对象
　　public MediaController myMediaController；//定义公有MediaController对象
　　private File file；//定义私有File对象
　　public void onCreate(Bundle savedInstanceState) {//子类重写基类onCreate方法
　　　　super.onCreate(savedInstanceState)；//调用基类onCreate方法
　　　　setContentView(R.layout.*main*)；//设置屏幕布局
　　　　myVideoView=(VideoView)findViewById(R.id.*videoView*)；//查找VideoView对象
　　　　myMediaController=**new** MediaController(**this**)；//创建MediaController对象
　　　　file=**new** File("/mnt/sdcard/video/football.3gp")；//创建File对象
　　　　getWindow().setFormat(PixelFormat.*TRANSLUCENT*)；//设置窗体透明风格
　　　　if(file.exists()){ //检查文件是否存在
　　　　//设置VideoView与mController建立联系
　　　　 myVideoView.setVideoPath(file.getAbsolutePath())；//设置视频文件绝对路径
　　　　//让VideoView　获取焦点。
　　　　myVideoView.setMediaController(myMediaController);
　　　　 myVideoView.requestFocus();
　　　　} **else**{
　　　　　　Toast.*makeText*(CH6_2.**this**,"文件不存在",

```
            Toast.LENGTH_LONG).show();
        }
    }
}
```

6.3.5 知识点 5：录音

在上一节中我们学习了如何来用 Android 播放音乐文件，本节将来学习如何用 Android 手机录制一段音频文件。Android 系统提供了 MediaRecorder 类来实现音频和视频的录制。首先来看一下 Android 的 MediaRecorder 为我们提供了哪些方法。以一个录音为实例，该实例运行如图 6-6 所示。

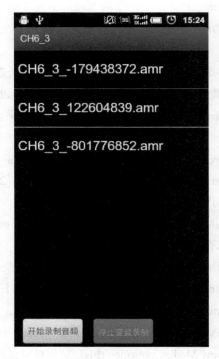

图 6-6 录制音频

具体使用 MediaRecorder 的方法可以归纳为如下几步。

（1）创建一个 android.media.MediaRecorder 的新实例。

（2）使用 MediaRecorder.setAudioSource()设置音频源，一般要使用 MediaRecorder.AudioSource.MIC。

（3）使用 MediaRecorder.setOutputFormat()设置输出文件的格式。

（4）使用 MediaRecorder.setOutputFile()设置输出文件的名字。

（5）使用 MediaRecorder.setAudioEncoder()设置音频编码。

（6）调用 MediaRecorder 实例的 MediaRecorder.prepare()。

（7）MediaRecorder.start()开始获取音频。

(8) 调用 MediaRecorder.stop().停止。

(9) 用完 MediaRecorder 实例后，调用 MediaRecorder.release()，释放资源。

在这个例子中，我们通过一个列表将所录制的音频文件进行罗列，同时通过两个按钮来控制对音频的录制和播放。布局文件如代码清单 6-8 所示。

代码清单 6-8 第 6 章\CH6_3\res\layout\main.xml

```xml
<?xml version="1.0" encoding="utf-8"?>
<RelativeLayout xmlns:android="http://schemas.android.com/apk/res/android"
    xmlns:tools="http://schemas.android.com/tools"
    android:layout_width="match_parent"
    android:layout_height="match_parent" >
    <ListView
 android:id="@id/android:list"
        android:layout_width="match_parent"
          android:layout_height="wrap_content"
        android:layout_weight="1"
        android:drawSelectorOnTop="false" />
    <LinearLayout
        android:layout_width="fill_parent"
        android:layout_height="wrap_content"
        android:layout_alignParentBottom="true"
        android:layout_alignParentLeft="true"
        android:layout_alignParentRight="true"
        android:layout_marginRight="10dp"
        android:layout_marginLeft="10dp" >
    <Button
 android:id="@+id/mediarecorder1_AudioStartBtn"
        android:layout_width="wrap_content"
        android:layout_height="wrap_content"
        android:text="开始录制音频" />
    <Button
 android:id="@+id/mediarecorder1_AudioStopBtn"
        android:layout_width="wrap_content"
        android:layout_height="wrap_content"
      android:layout_marginLeft="10dp"
        android:text="停止音频录制" />
        </LinearLayout>
</RelativeLayout>
```

在主程序中需要注意两点，第一，是列表需要对卡上的文件进行筛选，将录音文件筛选出来；第二，当单击列表中的某一项时需要播放文件。主程序如代码清单 6-9 所示。

代码清单 6-9 第 6 章\CH6_3\src\sziit\practice\chapter6\CH6_3.java

```java
public class CH6_3 extends ListActivity {//从ListActivity基类派生子类CH6_3
    public Button btnStartRecord; //定义公有Button对象
    public Button btnStopRecord; //定义公有Button对象
    public File audioFile; // 定义录制的音频文件File对象
    public File filePath; //音频文件路径
    private MediaRecorder mMediaRecorder; //定义私有MediaRecorder对象
    private List<String> mMusicList = new ArrayList<String>(); // 定义私有录音文件列表对象
    private String strTempFile = "CH6_3_"; // 定义私有临时文件的前缀
    public void onCreate(Bundle savedInstanceState) {//子类重写基类onCreate方法
        super.onCreate(savedInstanceState); //调用基类onCreate方法
        setContentView(R.layout.main); //设置屏幕布局
        btnStartRecord=(Button) findViewById(R.id.mediarecorder1_AudioStartBtn); //查找Button
        btnStopRecord= (Button) findViewById(R.id.mediarecorder1_AudioStopBtn);
        btnStartRecord.setEnabled(true); //使能btnStartRecord对象
        btnStopRecord.setEnabled(false); //禁止btnStopRecord对象
        if (Environment.getExternalStorageState().equals(android.os.Environment.MEDIA_MOUNTED))
        {
            filePath = Environment.getExternalStorageDirectory(); // 得到SD卡得路径
            musicList(); // 更新所有录音文件到List中
        } else
        {   //显示错误信息
            Toast.makeText(CH6_3.this, "没有SD卡", Toast.LENGTH_LONG).show();
        }
        btnStartRecord.setOnClickListener(new Button.OnClickListener(){//设置监听器
            public void onClick(View arg0) {//处理回调方法onClick
                try
                {
                    //实例化MediaRecorder对象
                    mMediaRecorder = new MediaRecorder();
mMediaRecorder.setAudioSource(MediaRecorder.AudioSource.MIC); //设置麦克风
mMediaRecorder.setOutputFormat(MediaRecorder.OutputFormat.DEFAULT);
                    //设置音频文件的编码：AAC/AMR_NB/AMR_MB/Default
mMediaRecorder.setAudioEncoder(MediaRecorder.AudioEncoder.DEFAULT);
```

///设置输出文件的路径
try
{///创建临时音频文件
  audioFile = File.*createTempFile*(strTempFile, ".amr", filePath);
} **catch** (Exception e)//捕获异常
{
    e.printStackTrace(); //显示异常信息
}
mMediaRecorder.setOutputFile(audioFile.getAbsolutePath()); //设置输出路径
mMediaRecorder.prepare(); //准备
mMediaRecorder.start(); //开始
//按钮状态
btnStartRecord.setEnabled(**false**); //禁止btnStartRecord按钮
btnStopRecord.setEnabled(**true**); //使能btnStopRecord按钮
} **catch** (IOException e) //捕获异常
{
    e.printStackTrace(); //显示异常信息
}
}
});
btnStopRecord.setOnClickListener(**new** Button.OnClickListener(){//设置单击监听器
    public void onClick(View v) {//处理单击回调方法onClick
        if (audioFile != **null**)
        {
            mMediaRecorder.stop(); //停止录音
            mMusicList.add(audioFile.getName()); //将录音文件添加到List中
            ArrayAdapter<String> musicList = **new** ArrayAdapter<String>(CH6_3.**this**,
                android.R.layout.*simple_list_item_1*, mMusicList);
            setListAdapter(musicList);
            mMediaRecorder.release(); //释放MediaRecorder
            mMediaRecorder = **null**;
            btnStartRecord.setEnabled(**true**); // 按钮状态
            btnStopRecord.setEnabled(**false**);
        }
    }
});

```java
        }
        //播放录音文件
        private void playMusic(File file)
        {
            Intent intent = new Intent();    //创建意图(Intent)对象
            intent.addFlags(Intent.FLAG_ACTIVITY_NEW_TASK);    //设置意图创建的活动任务类型标记
            intent.setAction(android.content.Intent.ACTION_VIEW);    //设置意图执行的动作
            intent.setDataAndType(Uri.fromFile(file), "audio");    //设置文件类型
            startActivity(intent);    //启动意图切换到新的活动(Activity)
        }
        //当我们单击列表时，播放被单击的音乐
        protected void onListItemClick(ListView l, View v, int position, long id)
        {
            File playfile = new File(filePath.getAbsolutePath() + File.separator
                    + mMusicList.get(position));    //得到被单击的文件
            playMusic(playfile);    // 播放
        }
        //播放列表
        public void musicList()
        {
            // 取得指定位置的文件设置显示到播放列表
            File home = filePath;    //目录路径
            if (home.listFiles(new MusicFilter()).length > 0)
            {
                for (File file : home.listFiles(new MusicFilter()))
                {//遍历文件目录
                    mMusicList.add(file.getName());
                }
                ArrayAdapter<String> musicList = new ArrayAdapter<String>(CH6_3.this,
                        android.R.layout.simple_list_item_1, mMusicList);
                setListAdapter(musicList);
            }
        }
    }
    //过滤文件类型
    class MusicFilter implements FilenameFilter
    {
```

```
public boolean accept(File dir, String name)
{
    return (name.endsWith(".amr"));
}
}
```

6.4 项目实施

在掌握了上面的音频播放的基本知识以后，下面就来看一下在本章开始之前提到的多媒体播放功能的实现。首先看一下布局文件，如代码 6-10 所示。

代码清单 6-10 第 6 章\CH6_3\src\sziit\practice\chapter6\CH6_3.java

```xml
<?xml version="1.0" encoding="UTF-8"?>
<RelativeLayout xmlns:android="http://schemas.android.com/apk/res/android"
    android:layout_width="fill_parent"
    android:layout_height="fill_parent"
    android:background="@color/white"
    android:orientation="vertical" >
    <LinearLayout
        android:id="@+id/titleLayout"
        android:layout_width="fill_parent"
        android:layout_height="wrap_content" >
        <!—标题栏布局 -->
        <include layout="@layout/comm_title_bar" />
    </LinearLayout>
    <!-- tab 选项 -->
    <LinearLayout
        android:id="@+id/commoditysort_product_tab_layout"
        android:layout_width="fill_parent"
        android:layout_height="wrap_content"
        android:layout_below="@+id/titleLayout"
        android:layout_marginRight="10dip"
        android:gravity="center"
        android:orientation="horizontal" >
        <TextView
            android:id="@+id/commoditysort_product_lefttab_btn"
            android:layout_width="fill_parent"
            android:layout_height="35dip"
            android:layout_marginLeft="10dip"
```

```xml
            android:layout_weight="1"
            android:background="@drawable/detail_tab_background"
            android:gravity="center"
            android:text="@string/guide_information"
            android:textColor="@color/detail_tab_color"
            android:textSize="@dimen/sort_btn_textsize" />
        <TextView
            android:id="@+id/commoditysort_product_righttab_btn"
            android:layout_width="fill_parent"
            android:layout_height="35dip"
            android:layout_weight="1"
            android:background="@drawable/detail_tab_background"
            android:gravity="center"
            android:text="@string/detail_information"
            android:textColor="@color/detail_tab_color"
            android:textSize="@dimen/sort_btn_textsize" />
    </LinearLayout>
    <LinearLayout
        android:layout_width="fill_parent"
        android:layout_height="fill_parent"
        android:layout_above="@+id/tool_bar"
        android:layout_below="@+id/commoditysort_product_tab_layout"
        android:orientation="vertical" >
        <com.uxzv.leyou.widget.ObservableScrollView
            android:id="@+id/scrollView"
            android:layout_width="fill_parent"
            android:layout_height="fill_parent"
            android:layout_weight="1.0"
            android:fillViewport="true"
            android:scrollbars="vertical" >
            <LinearLayout
                android:id="@+id/lite_list"
                android:layout_width="fill_parent"
                android:layout_height="fill_parent"
                android:background="@color/white" >
            </LinearLayout>
        </com.uxzv.leyou.widget.ObservableScrollView>
```

```xml
</LinearLayout>
<LinearLayout
    android:id="@+id/popup_more"
    android:layout_width="wrap_content"
    android:layout_height="wrap_content"
    android:layout_above="@+id/tool_bar"
    android:layout_alignParentRight="true"
    android:layout_gravity="bottom|right|center"
    android:background="@drawable/popup"
    android:gravity="center"
    android:orientation="vertical"
    android:visibility="gone" >
    <TextView
        android:id="@+id/share_sina"
        android:layout_width="wrap_content"
        android:layout_height="wrap_content"
        android:layout_marginTop="2.0dip"
        android:background="@drawable/popup_option"
        android:clickable="true"
        android:paddingLeft="10.0dip"
        android:paddingRight="10.0dip"
        android:paddingTop="7.0dip"
        android:text="@string/lacl_tour_bottom_share_1"
        android:textAppearance="?android:textAppearanceMediumInverse"
        android:textColor="@color/white" />
    <View
        android:layout_width="fill_parent"
        android:layout_height="1.0dip"
        android:background="@drawable/toolbar_menu_line" />
    <TextView
        android:id="@+id/share_tencent"
        android:layout_width="wrap_content"
        android:layout_height="wrap_content"
        android:layout_marginBottom="10dip"
        android:background="@drawable/popup_option"
        android:clickable="true"
        android:paddingLeft="10.0dip"
```

```
            android:paddingRight="10.0dip"
            android:paddingTop="5.0dip"
            android:text="@string/lacl_tour_bottom_share_2"
            android:textAppearance="?android:textAppearanceMediumInverse"
            android:textColor="@color/white" />
</LinearLayout>
    <LinearLayout
        android:id="@+id/tool_bar"
        android:layout_width="fill_parent"
        android:layout_height="55.0dip"
        android:layout_alignParentBottom="true"
        android:layout_gravity="bottom"
        android:background="@drawable/tool_bar"
        android:gravity="center_vertical"
        android:orientation="horizontal" >
        <Button
            android:id="@+id/collected"
            android:layout_width="0.0dip"
            android:layout_height="fill_parent"
            android:layout_weight="1.0"
            android:background="@drawable/imagebutton_background"
            android:drawableTop="@drawable/toolbar_collect_add"
            android:paddingTop="10.0dip"
            android:text="@string/local_tour_bottom_tab1"
            android:textColor="@color/white"
            android:textSize="10.0sp" />
        <Button
            android:id="@+id/comment"
            android:layout_width="0.0dip"
            android:layout_height="fill_parent"
            android:layout_weight="1.0"
            android:background="@drawable/imagebutton_background"
            android:drawableTop="@drawable/toolbar_comment"
            android:paddingTop="10.0dip"
            android:text="看看点评"
            android:textColor="@color/white"
            android:textSize="10.0sp"
```

```
                    android:visibility="gone" />
                <Button
                    android:id="@+id/share"
                    android:layout_width="0.0dip"
                    android:layout_height="fill_parent"
                    android:layout_weight="1.0"
                    android:background="@drawable/imagebutton_background"
                    android:drawableTop="@drawable/toolbar_more"
                    android:paddingTop="10.0dip"
                    android:text="@string/local_tour_bottom_tab6"
                    android:textColor="@color/white"
                    android:textSize="10.0sp" />
        </LinearLayout>
</RelativeLayout>
```

具体看一下程序的实现，功能详细的代码可以参考本书项目源代码中包 com.uxzv.leyou.details 下的 LocalTourDetailActivity.java。在音频播放部分我们剪出来一起分析。如下：

```
//初始化播放器
public void initPlayer() {
        mediaPlayer = new MediaPlayer();  //创建 MediaPlayer 对象
        mAudio_id = getAudioID(detailData.getName());  //获取景点名称
    }
//通过名字来分别获取音频的 ID
    private int getAudioID(String name) {
        if (name.equals("黄果树瀑布")) {
            audio.setVisibility(View.VISIBLE);  //设置 audio 的可见属性
            return R.raw.hgs1;  //返回播放的音频文件
        } else if (name.equals("龙宫")) {
            audio.setVisibility(View.VISIBLE);  //设置 audio 的可见属性
            return R.raw.hgs2;  //返回播放的音频文件
        } else if (name.equals("陡坡塘景区")) {
            audio.setVisibility(View.VISIBLE);  //设置 audio 的可见属性
            return R.raw.hgs3;  //返回播放的音频文件
        } else if (name.equals("黄果树奇石馆")) {
            audio.setVisibility(View.VISIBLE);  //设置 audio 的可见属性
            return R.raw.hgs5;  //返回播放的音频文件
        } else if (name.equals("石头寨景区")) {
            audio.setVisibility(View.VISIBLE);  //设置 audio 的可见属性
```

```java
                return R.raw.hgs6;  //返回播放的音频文件
            } else if (name.equals("郎宫")) {
                audio.setVisibility(View.VISIBLE);  //设置 audio 的可见属性
                return R.raw.hgs7;  //返回播放的音频文件
            } else if (name.equals("夜郎洞")) {
                audio.setVisibility(View.VISIBLE);  //设置 audio 的可见属性
                return R.raw.hgs8;  //返回播放的音频文件
            } else if (name.equals("黄果树漂流")) {
                audio.setVisibility(View.VISIBLE);  //设置 audio 的可见属性
                return R.raw.hgs9;  //返回播放的音频文件
            } else if (name.equals("滴水潭")) {
                audio.setVisibility(View.VISIBLE);  //设置 audio 的可见属性
                return R.raw.hgs10;  //返回播放的音频文件
            } else if (name.equals("红岩天书")) {
                audio.setVisibility(View.VISIBLE);  //设置 audio 的可见属性
                return R.raw.hgs11;  //返回播放的音频文件
            } else if (name.equals("天星桥景区")) {
                audio.setVisibility(View.VISIBLE);  //设置 audio 的可见属性
                return R.raw.hgs12;  //返回播放的音频文件
            } else if (name.equals("织金洞")) {
                audio.setVisibility(View.VISIBLE);  //设置 audio 的可见属性
                return R.raw.hgs13;  //返回播放的音频文件
            } else if (name.equals("螺丝滩瀑布")) {
                audio.setVisibility(View.VISIBLE);  //设置 audio 的可见属性
                return R.raw.hgs20;  //返回播放的音频文件
            } else if (name.equals("坝岭河大桥")) {
                audio.setVisibility(View.VISIBLE);  //设置 audio 的可见属性
                return R.raw.hgs21;  //返回播放的音频文件
            }
            return R.raw.hgs1;  //返回播放的音频文件
        }
    private boolean isMediaPause;
        private boolean mNeetPause;
        protected void onPause() {//重写 onPause 方法
            if (mediaPlayer != null && mediaPlayer.isPlaying() && mNeetPause) {
                mediaPlayer.pause();  //停止播放
                audio.setImageResource(R.drawable.audio_normal);  //设置图像资源
```

```
            isMediaPause = true； //设置停止播放标记
            mNeetPause = false;
        }
        super.onPause()； //调用基类的 onPause 方法
    }
    protected void onResume() {//重写 onResume 方法
        if (mediaPlayer != null && isMediaPause) {
            mediaPlayer.start()； //开始播放
            audio.setImageResource(R.drawable.audio_pressed)； //设置图像资源
            isMediaPause = false;
        }
        super.onResume()； //调用基类的 onResume 方法
    }
    protected void onStop() {//重写 onStop 方法
        if (collectDBHelper != null) {
            collectDBHelper.close()； //关闭 collectDBHelper 对象
            collectDBHelper = null;
        }
        super.onStop()； //调用基类的 onStop 方法
    }
//根据用户所点按钮来进行相应操作，当单击音频播放按钮时操作如下
if (!isAudioPressed) {
                isAudioPressed = true;
                try {//创建 MediaPlayer 对象
                    mediaPlayer = MediaPlayer.create(
                        LocalTourDetailActivity.this, mAudio_id);
                } catch (IllegalArgumentException e) {//捕获 IllegalArgumentException 异常
                    e.printStackTrace()； //显示异常信息
                } catch (IllegalStateException e) {//捕获 IllegalStateException 异常
                    e.printStackTrace()； //显示异常信息
                }
                mediaPlayer.start()； //开始播放
                audio.setImageResource(R.drawable.audio_pressed)； //设置图像资源
                mediaPlayer.setOnCompletionListener(new OnCompletionListener() {
                    public void onCompletion(MediaPlayer mp) {//处理回调方法
                        audio.setImageResource(R.drawable.audio_normal);
                        isAudioPressed = false;
```

```
                    }
                });
            } else {
                isAudioPressed = false;
                if (mediaPlayer != null) {
                    mediaPlayer.stop();  //停止播放
                }
                audio.setImageResource(R.drawable.audio_normal);  //设置图像资源
            }
```

6.5 技术拓展

自从苹果公司在 2007 年发布第一代 iPhone 以来，以前看似和手机挨不着边的传感器也逐渐成为手机硬件的重要组成部分。如果读者使用过 iPhone、HTC Dream、HTC Magic、HTC Hero 以及其他 Android 手机，会发现通过将手机横向或纵向放置，屏幕会随着手机位置的不同而改变方向。这种功能就需要通过重力传感器来实现，除了重力传感器，还有很多其他类型的传感器被应用到手机中，如磁阻传感器就是最重要的一种传感器。虽然手机可以通过 GPS 来判断方向，但在 GPS 信号不好或根本没有 GPS 信号的情况下，GPS 就形同虚设。这时通过磁阻传感器就可以很容易判断方向（东、南、西、北）。有了磁阻传感器，也使罗盘（俗称指向针）的电子化成为可能。下面就来详细学习一下 Android 对传感器的支持。

在 Android2.3 gingerbread 系统中，google 提供了 11 种传感器供应用层使用，如下所示。

1．加速度传感器（SENSOR_TYPE_ACCELEROMETER）

加速度传感器又叫 G-sensor，返回 x、y、z 三轴的加速度数值。该数值包含地心引力的影响，单位是 m/s^2。将手机平放在桌面上，x 轴默认为 0，y 轴默认为 0，z 轴默认为 9.81。将手机朝下放在桌面上，z 轴为 -9.81；将手机向左倾斜，x 轴为正值；将手机向右倾斜，x 轴为负值；将手机向上倾斜，y 轴为负值；将手机向下倾斜，y 轴为正值。

加速度传感器可能是最为成熟的一种 mems 产品，市场上的加速度传感器种类很多。手机中常用的加速度传感器有 BOSCH（博世）的 BMA 系列、AMK 的 897X 系列、ST 的 LIS3X 系列等。这些传感器一般提供±2G 至±16G 的加速度测量范围，采用 I2C 或 SPI 接口和 MCU 相连，数据精度小于 16bit。

2．磁力传感器（SENSOR_TYPE_MAGNETIC_FIELD）

磁力传感器简称为 M-sensor，返回 x、y、z 三轴的环境磁场数据。该数值的单位是微特斯拉（micro-Tesla），用 uT 表示，单位也可以是高斯（Gauss），1Tesla=10000Gauss。

硬件上一般没有独立的磁力传感器，磁力数据由电子罗盘传感器提供（E-compass）。电子罗盘传感器同时提供下文的方向传感器数据。

3．方向传感器（SENSOR_TYPE_ORIENTATION ）

方向传感器简称为 O-sensor，返回三轴的角度数据，方向数据的单位是角度。为了得到精确的角度数据，E-compass 需要获取 G-sensor 的数据，经过计算生产 O-sensor 数据，否则只能获取水平方向的角度。方向传感器提供 3 个数据，分别为 azimuth、pitch 和 roll。

- azimuth：方位，返回水平时磁北极和 y 轴的夹角，范围为 0°～360°。0°=北，90°=东，180°=南，270°=西。
- pitch：x 轴和水平面的夹角，范围为–180°～180°。当 z 轴向 y 轴转动时，角度为正值。
- roll：y 轴和水平面的夹角，由于历史原因，范围为–90°～90°。当 x 轴向 z 轴移动时，角度为正值。

电子罗盘在获取正确的数据前需要进行校准，通常可用 8 字校准法。8 字校准法要求用户使用需要校准的设备在空中做 8 字晃动，原则上尽量多地让设备法线方向指向空间的所有 8 个象限。手机中使用的电子罗盘芯片有 AKM 公司的 897X 系列、ST 公司的 LSM 系列以及雅马哈公司等。

由于需要读取 G-sensor 数据并计算出 M-sensor 和 O-sensor 数据，因此厂商一般会提供一个后台 daemon 来完成工作，电子罗盘算法一般是公司私有产权。

4．陀螺仪传感器（SENSOR_TYPE_GYROSCOPE）

陀螺仪传感器又叫做 Gyro-sensor，返回 x、y、z 三轴的角加速度数据。角加速度的单位是 radians/second。根据 Nexus S 手机实测：

- 水平逆时针旋转，z 轴为正；
- 水平逆时针旋转，z 轴为负；
- 向左旋转，y 轴为负；
- 向右旋转，y 轴为正；
- 向上旋转，x 轴为负；
- 向下旋转，x 轴为正。

ST 的 L3G 系列的陀螺仪传感器比较流行，iPhone4 和 Google 的 nexus s 中使用该种传感器。

5．光线感应传感器（SENSOR_TYPE_LIGHT）

光线感应传感器检测实时的光线强度，光强单位是 lux，其物理意义是照射到单位面积上的光通量。光线感应传感器主要用于 Android 系统的 LCD 自动亮度功能。可以根据采样到的光强数值实时调整 LCD 的亮度。

6．压力传感器（SENSOR_TYPE_PRESSURE）

压力传感器返回当前的压强，单位是百帕斯卡 hectopascal（hPa）。

7．温度传感器（SENSOR_TYPE_TEMPERATURE）

温度传感器返回当前的温度。

8．接近传感器（SENSOR_TYPE_PROXIMITY）

接近传感器检测物体与手机的距离，单位是厘米。一些接近传感器只能返回远和近两个状态，因此，接近传感器将最大距离返回远状态，小于最大距离返回近状态。接近传感器可用于接听电话时自动关闭 LCD 屏幕以节省电量。一些芯片集成了接近传感器和光线传感器两者功能。

9．重力传感器（SENSOR_TYPE_GRAVITY）

重力传感器简称 GV-sensor，输出重力数据。在地球上，重力数值为 9.8，单位是 m/s^2。坐标系统与加速度传感器相同。当设备复位时，重力传感器的输出与加速度传感器相同。

10．线性加速度传感器（SENSOR_TYPE_LINEAR_ACCELERATION 10）

线性加速度传感器简称 LA-sensor。线性加速度传感器是加速度传感器减去重力影响获取的数据。单位是 m/s²，坐标系统与加速度传感器相同。加速度传感器、重力传感器和线性加速度传感器的计算公式如下：

$$加速度 = 重力 + 线性加速度$$

11．旋转矢量传感器（SENSOR_TYPE_ROTATION_VECTOR）

旋转矢量传感器简称 RV-sensor。旋转矢量代表设备的方向，是一个将坐标轴和角度混合计算得到的数据。RV-sensor 输出 3 个数据：

- x*sin(theta/2)；
- y*sin(theta/2)；
- z*sin(theta/2)。

sin(theta/2)是 RV 的数量级；RV 的方向与轴旋转的方向相同；RV 的 3 个数值，与 cos(theta/2) 组成一个四元组。RV 的数据没有单位，使用的坐标系与加速度相同。GV、LA 和 RV 的数值没有物理传感器可以直接给出，需要 G-sensor、O-sensor 和 Gyro-sensor 经过算法计算后得出。算法一般是传感器公司的私有产权。

我们在做传感器的开发时大致可以分为如下几步。

- 获取传感器服务：sm =(SensorManager) getSystemService(SENSOR_SERVICE)。
- 获取指定类型的传感器 sm.getDefaultSensor(int type)。
- 一般在 Activity 的 onResume()方法中使用 SensorManger 的 registerListener()为指定的传感器注册监听器即可，如下所示：

sm.registerListener(listener, s1, SensorManager.SENSOR_DELAY_ GAME)；//最后一个参数为监听频率

在 Listener 中的 SensorEvent 对象的 values 方法返回 float[]，包含不同传感器返回的数据值，对于不同的传感器，返回的值的个数是不一样的。下面就通过一个实例来观察手机中所包含的传感器类型，代码如代码清单 6-11 所示。

代码清单 6-11 第 6 章\CH6_4\src\sziit\practice\chapter6\CH6_4.java
public class CH6_4 extends Activity {//从extends基类派生子类CH6_4
    private TextView show；//定义私有TextView对象
    private SensorManager sm；//定义私有创感器管理器(SensorManager)对象
    private StringBuffer str；//定义私有字符串缓冲器(StringBuffer)对象
    private List<Sensor> allSensors；//定义私有列表容器
    private Sensor s；//定义私有创感器(Sensor)对象
    public void onCreate(Bundle savedInstanceState) {//子类重写基类onCreate方法
        super.onCreate(savedInstanceState)；//调用基类onCreate方法
        setContentView(R.layout.main)；//设置屏幕布局
        show = (TextView) findViewById(R.id.show)；//从布局文件中查找TextView对象
        sm = (SensorManager) getSystemService(Context.SENSOR_SERVICE)；//获得系统服务

```
        allSensors = sm.getSensorList(Sensor.TYPE_ALL); // 获得传感器列表
        getAllSensor();
    }
    private void getAllSensor() {//实现私有方法getAllSensor
        str = new StringBuffer(); //创建StringBuffer对象
        str.append("该手机有" + allSensors.size() + "个传感器,分别是:\n");
        for (int i = 0; i < allSensors.size(); i++) {
            s = allSensors.get(i);
            str.append("设备名称:" + s.getName() + "\n");
            str.append("设备版本:" + s.getVersion() + "\n");
            str.append("通用类型号:" + s.getType() + "\n");
            str.append("设备商名称:" + s.getVendor() + "\n");
            str.append("传感器功耗:" + s.getPower() + "\n");
            str.append("传感器分辨率:" + s.getResolution() + "\n");
            str.append("传感器最大量程:" + s.getMaximumRange() + "\n");
            switch (s.getType()) {//处理不同类型的创感器
            case Sensor.TYPE_ACCELEROMETER:
                str.append("传感器类型 "+ "加速度传感器\n");
                break;
            case Sensor.TYPE_GYROSCOPE:
                str.append("传感器类型 " + "陀螺仪传感器\n");
                break;
            case Sensor.TYPE_LIGHT:
                str.append("传感器类型 " + "环境光线传感器\n");
                break;
            case Sensor.TYPE_MAGNETIC_FIELD:
                str.append("传感器类型 " + "电磁场传感器\n");
                break;
            case Sensor.TYPE_ORIENTATION:
                str.append("传感器类型 " + "方向传感器\n");
                break;
            case Sensor.TYPE_PRESSURE:
                str.append("传感器类型 " + "压力传感器\n");
                break;
            case Sensor.TYPE_PROXIMITY:
                str.append("传感器类型 " + "距离传感器\n");
                break;
```

```
            case Sensor.TYPE_TEMPERATURE:
                str.append("传感器类型  " + "温度传感器\n");
                break;
            default:
                str.append("传感器类型  " + "未知传感器\n");
                break;
            }
        }
        show.setText(str);  //显示相关信息
    }
}
```

6.6 本章小结

本章中我们着重介绍了 Android 系统的多媒体知识，包括网络访问、服务、音频播放、视频播放以及录音实现。通过这章的学习让读者了解 Android 系统所提供的丰富的多媒体内容。我们通过 MediaPlayer 来进行音乐的播放，借助 VideoView 来播放视频文件；在音频的录制上我们着重介绍了 MediaRecorder 类的使用，通过 MediaRecorder 能够录制音频文件，同样，也可以使用 MediaRecorder 来进行视频的录制，读者可以在音频录制的基础上进行扩展，学习视频的录制。

最后我们在技术拓展部分讲解了 Android 对传感器的支持，以求通过借助更多的传感器，让手机更加智能化，让生活更加方便。

6.7 强化练习

1. 简述 Android 网络连接的不同方式。
2. Android Service 与 Activity 的区别是什么？
3. Android 中 Service 的启动方式有_____和_____两种方式。
4. 使用 Android MediaPlayer 进行音频播放时音频文件的加载方式有哪些？
5. Android 封装了一个视频播放的控件，可以直接进行视频的播放，该视频控件的名称是_____。
6. 简述 MediaRecorder 的使用步骤。
7. 列举 3 种常用的手机传感器。

第7章 综合实训1 手机乐游项目

7.1 项目分析

在本书最后，我们通过具体的项目真正地锻炼学生所学的 Android 知识。首先，要先确定我们的需求，只有了解需求后方能正确地按照预期进行开发。前期需求的确定是后面项目测试和验收的重要依据，因此，需要认真严肃地对待需求分析文档。下面就来看一下手机乐游客户端项目的需求分析文档，主要介绍需求文档中对项目的基本需求，具体内容如下。

（1）UR-F 0010　景区选择

需求描述：旅游者通过手机客户端软件，选择自己需要查看的景区，支持自动定位及手动选择两种方式。

① UR-F 0010-0011　自动定位选择

需求描述：本系统的手机客户端软件根据旅游者的位置信息，自动选择最近的景区，或者列出最近的景区供用户选择。

该功能要求旅游者的手机支持定位服务。

② UR-F 0010-0012　手动选择

需求描述：旅游者通过在手机客户端软件中输入旅游景区相关关键字，或者按照行政区域进行导航，查询需要查看的景区。

（2）UR-F 0020　景区信息浏览

需求描述：旅游者通过手机客户端软件浏览景区内的景点、美食、住宿、购物等信息；可以在浏览的过程中通过微博分享的方式来分享自己所感兴趣的信息；可以评论，并查看相关点评信息；对感兴趣的信息可以进行收藏；可以预订景区门票。除此之外，还包括一些实用信息，包括门票信息、重要电话、交通信息和注意事项等。具体描述如下。

① UR-F 0020-0021　景点信息

需求描述：显示导游信息和详细信息，包括文字、图片及音频资料。另外，还提供旅游者门票预定功能。

在该模块中，有危险求救功能，通过此功能可以让旅游者在游玩的过程中因迷路而使自己无法正常游玩或遭遇不可估计的情况下使用，获得旅游管理处的及时帮助，同时旅游管理者可以掌握景区实时动态；还有实时推送功能，通过这个功能，为旅游者提供当前位置周边的餐饮、住宿、土特产的信息推送服务；还可选择景点语音介绍等功能；还有可使旅游者在浏览中下订单等功能。

② UR-F 0020-0022   美食信息

需求描述：景区周边餐饮美食信息展示，并提供分类导航及根据关键字搜索功能。

③ UR-F 0020-0023   购物信息

需求描述：景区周边商家商品信息展示，并提供分类导航及根据关键字搜索功能。

④ UR-F 0020-0024   住宿信息

需求描述：景区周边酒店、农家乐信息展示，并提供分类导航及根据关键字搜索功能。

⑤ UR-F 0020-0025   实用信息

需求描述：景区实用信息展示，包括门票信息、重要电话、交通信息和注意事项等。

（3）UR-F 0030   景区导航

需求描述：旅游者通过手机客户端软件查看整个景区的地理分布情况，方便规划游玩路线和获取景区的相对位置，同时通过定位服务，为旅游者在景区内进行游览导航，并播放当前位置景点的图文、音频解说。

① UR-F 0030-0031   静态导游图

需求描述：静态导游地图展示，展示主要景点及道路。

② UR-F 0030-0032   实时景区导航

需求描述：通过定位服务，为旅游者在景区内进行游览导航，为旅游者播放当前位置景点的图文、音频解说等。

（4）UR-F 0040   旅游资源管理

需求描述：旅游管理部门、系统管理者以及商户可以对旅游资源信息进行维护。

① UR-F 0040-0041   景区管理

需求描述：旅游管理部门、系统管理者可以通过 Web 浏览器，维护旅游云平台中的景区信息，包括增加、删除、修改、查看景区信息。

② UR-F 0040-0042   景点管理

需求描述：旅游管理部门、系统管理者可以通过 Web 浏览器，维护旅游云平台中某景区的景点信息，包括增加、删除、修改、查看景点信息。

③ UR-F 0040-0043   商家信息管理

需求描述：旅游管理部门、系统管理者可以通过 Web 浏览器，维护旅游云平台中的商家信息，包括增加、删除、修改、查看商家信息。

④ UR-F 0040-0044   商家商品管理

需求描述：旅游管理部门、系统管理者、商家可以通过 Web 浏览器，维护旅游云平台中的商家商品信息，包括增加、删除、修改、查看商家商品信息。

除以上对项目的基本需求外，还有如下一些其他需求，如下所示。

（1）UR-O 0010   手机端应用要求

需求描述：旅游者在使用该软件时，要求程序包大小小于 5MB（不含更新包及网络下载内容）；

程序响应时间少于 3s（典型主流智能手机在网络状态良好情况下）。

（2）UR-O 0020　Web 端应用要求

需求描述：旅游者户发出订单后，Web 客户端获得客户订单信息时间小于 3s（典型主流智能手机在网络状态良好情况下）；Web 客户端确认订单后，确认信息到达客户手机端的时间小于 3s（典型主流智能手机在网络状态良好情况下）。

（3）UR-O 0030　服务端应用要求

需求描述：旅游管理部门要求服务器在使用过程中 CPU 占用率<50%（典型主流服务器配置），并发用户支持个数。

（4）UR-O 0040　数据安全性要求

需求描述：在同一时间上最多能并发多少客户端的请求，用于传送数据，保证数据存储的安全。

（5）UR-O 0050　性能指标

需求描述：分析客户端的最大连接量和服务器端的最大并发量。

7.2　项目设计

以上内容对需求进行了归纳和整理，按照软件工程的思想，下面将根据需要需求分析对项目进行概要设计和详细设计。在这个阶段中所形成的文档和成果是为接下来的编码做准备，能够指导程序员进行程序开发。

在项目的概要设计阶段以流程图为主，详细描述整个系统的流程。项目的流程图如图 7-1 所示。

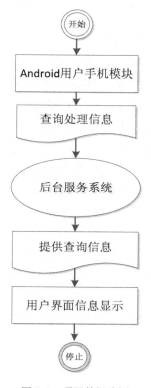

图 7-1　顶层数据流图

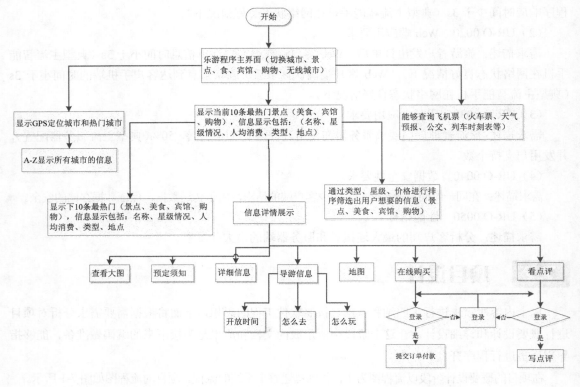

图 7-2 系统详细流程图

（1）主界面需求分析：在主界面中包括"当地游"、"景区地图"、"实用信息"、"订单管理"和"我的收藏" 5 个功能模块。

（2）当地游需求分析：当地游主要介绍当地城市的景点、美食、住宿和购物信息。用户可以单击景点、美食、宾馆、购物进入不同的界面，界面的布局尽量统一；界面显示要求：当前城市、排序操作按钮（包括类型、星级、价位）、数据量显示不要过大（暂定 10 条）；每条信息显示应该将最重要的信息简要显示出来，包括名称、人均消费、地点、类型、星级（可考虑 GPS 定位距离）情况；用户可单击该条信息进入详情页。其流程如 7-3 所示。

（3）详情页分为三部分组成：产品介绍、详细信息、预定须知。产品介绍主要显示星级情况、价位、类型、联系电话、景区简介、游览线路、指南（开放时间、优惠信息、旅游指南、景区交通）。

（4）景区详细地图显示，将该景区中的各个景点、饭店、停车场和公共厕所等信息显示在地图上，同时显示个人的位置，能够搜索个人附近的景点、饭店、停车场和公共厕所。单击地图上的标签可以显示该点的信息。

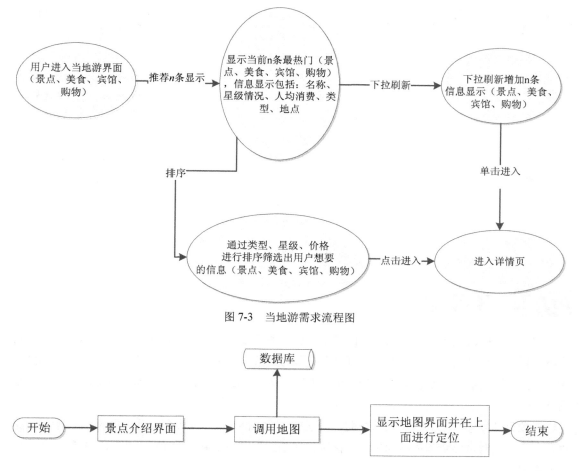

图 7-3 当地游需求流程图

图 7-4 景区详细地图显示

(5) 购物需求分析。

① 购物入口：景点、美食、住宿、购物详情界面提供支付入口，在支付流程中，必须确定已经有账号登录，以便记录用户购买信息。

② 购物订单信息界面：需提供产品名称、单价、数量、总价，提供购买者的手机号码。

③ 信息确认界面：当用户提交订单后，需要用户核实自己填写的信息，除用户填写的信息外，还要选择支付方式，支付方式选用支付宝支付，可以跳转到支付宝网站支付或者支付宝客户端软件支付（用户需安装，提示用户下载安装）。

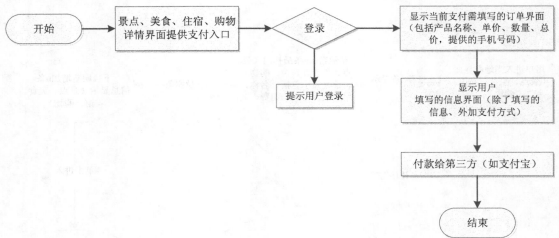

图 7-5　购物流程

7.3　项目实施

项目实施阶段则主要是借助前期的需求分析、概要设计、详细设计进行功能的实现。在编码阶段，项目组内部需要对各个功能进行划分，同时需要建立 svn 资源库（资源库建立本书不再详述）。每个成员需要分别完成自己的内容，同时需要对自己所负责的功能模块负责，并进行单元测试。

在项目开发中，还需要美工的协助。美工在项目中的主要职责是根据用户体验对程序的界面进行设计，同时需要将程序中的每个图标进行切割。程序员则利用美工设计的图标与界面进行功能的实现。

项目的包结果图如图 7-6 所示，其中各个包介绍如下：

- com.leyou.collect："我的收藏"模块处理代码；
- com.leyou.comment："评论"模块处理代码；
- com.leyou.detail："景点详情"页面处理代码；
- com.leyou.main："主界面"，启动界面入口页面处理代码；
- com.leyou.map："地图模式"模块处理代码；
- com.leyou.obj：程序公共变量、方法包；
- com.leyou.order：订单管理模块功能代码；
- com.leyou.pay："在线支付"功能模块代码；
- com.leyou.userinfo："用户信息"功能模块代码；
- com.leyou.util：程序工具包；
- com.leyou.weibo：微博分享模块功能代码。

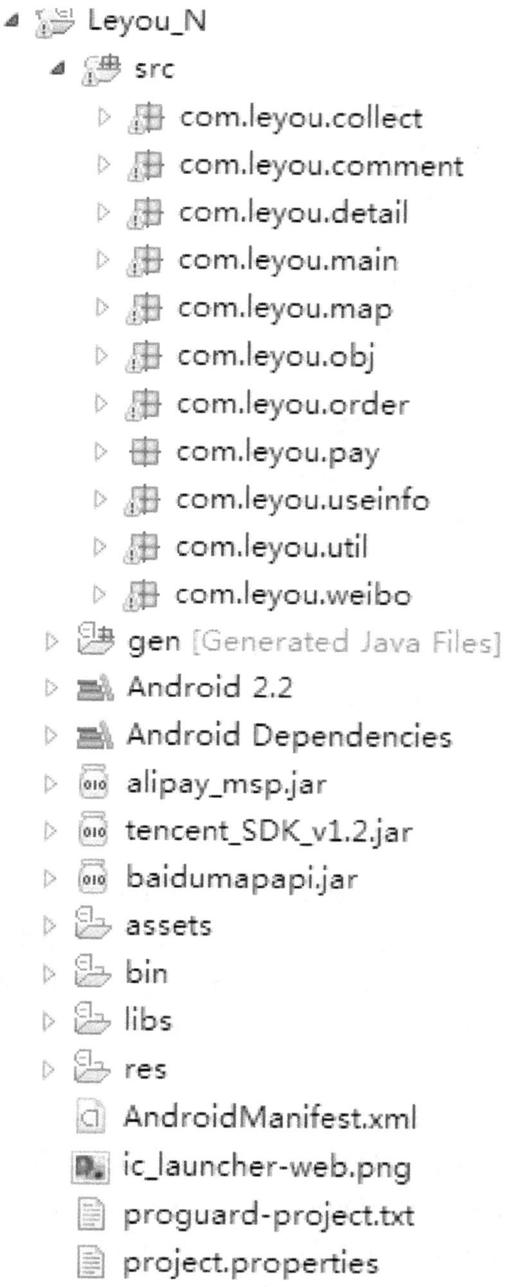

```
▲ Leyou_N
  ▲ src
    ▷ com.leyou.collect
    ▷ com.leyou.comment
    ▷ com.leyou.detail
    ▷ com.leyou.main
    ▷ com.leyou.map
    ▷ com.leyou.obj
    ▷ com.leyou.order
    ▷ com.leyou.pay
    ▷ com.leyou.useinfo
    ▷ com.leyou.util
    ▷ com.leyou.weibo
  ▷ gen [Generated Java Files]
  ▷ Android 2.2
  ▷ Android Dependencies
  ▷ alipay_msp.jar
  ▷ tencent_SDK_v1.2.jar
  ▷ baidumapapi.jar
  ▷ assets
  ▷ bin
  ▷ libs
  ▷ res
    AndroidManifest.xml
    ic_launcher-web.png
    proguard-project.txt
    project.properties
```

图 7-6 项目的包结果图

在程序中我们对腾讯微博、新浪微博、手机支付、百度地图进行了对接，故在程序中需要引用"tencent_sdk_v1.2.jar"、"baidumapapi.jar"、"alipay_map.jar"。同时为了丰富程序的动画效果，我们定义了自己的界面切换的效果，位于 res/anim 下，如图 7-7 所示。

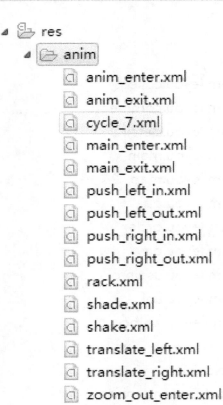

图 7-7　系统资源

通过前面的学习,我们已经对项目中的功能模块进行了讲解或联系,下面着重来讲解分享模块的实现。其中,功能代码位于 com.leyou.weibo 包下。需要注意的是,如果要将应用程序与新浪、腾讯等开放平台对接使用开放平台所提供的功能,如分享微博、分享到朋友圈等,就需要在其开放平台上进行注册和认证。具体可以参考新浪微博、腾讯微博官方要求。程序代码如下:

```
//发布按钮单击事件
private void initPublishBtn() {//微博功能实现
    mPublishWeibo_btn = (Button)findViewById(R.id.publishWeibo_btn);　//查找Button对象
    mPublishWeibo_btn.setOnClickListener(new View.OnClickListener() {//设置单击监听器
        public void onClick(View v) {//处理单击回调方法onClick
            if(!NetworkControl.isNetworkAvailable(getApplicationContext())) {//检测网络
                Toast.makeText(getApplicationContext(), "网络异常,请开启网络后再进行分享.", Toast.LENGTH_SHORT).show();　//显示网络异常信息
                return;
            }
            mContent = mWeiboShareContent_et.getText().toString();　//获得微博信息
            if(mWeiboType == WEIBOTYPE_TENCENT) {
```

```
                    loadTencentWeibo();  //加载腾讯微博
                }
                else if(mWeiboType == WEIBOTYPE_SINA) {
                    loadSinaWeibo();  //加载新浪微博
                }
            }
        });
}
    通过loadSinaWeibo()方法加载新浪微博发布界面。
private void loadSinaWeibo() {
    Weibo weibo = Weibo.getInstance();  //获得微博对象实例
    SharedPreferences shared = getSharedPreferences("myoauth", MODE_WORLD_READABLE);
    boolean haveOath = shared.getBoolean(SINA_AUTHENTICATION, false);
    if(haveOath) {
        String token = shared.getString(SINA_ACCESS_TOKEN, "");
        String expires_in = shared.getString(SINA_EXPIRES_IN, "0");
        AccessToken accessToken = new AccessToken(token, SINA_CONSUMER_SECRET);
        accessToken.setExpiresIn(expires_in);
        Weibo.getInstance().setAccessToken(accessToken);
        try {
            share2Sinaweibo(mWeiboShareContent_et.getText().toString(), null);
            Intent i = new Intent(WeiboShareActivity.this, ShareActivity.class);
            WeiboShareActivity.this.startActivity(i);
        } catch (WeiboException e) {
            e.printStackTrace();
            Toast.makeText(getApplicationContext(),
                    "share2Sinaweibo exception : " + e.getMessage(), Toast.LENGTH_LONG)
                    .show();
            Editor editor = shared.edit();
            editor.putBoolean(SINA_AUTHENTICATION, false);
            editor.commit();
        }
    }
    else {
        weibo.setupConsumerConfig(SINA_CONSUMER_KEY, SINA_CONSUMER_SECRET);
        weibo.setRedirectUrl(SINA_REDIRECT_URL);
```

```
        weibo.authorize(this, new AuthDialogListener());
    }
}
//分享信息到sina微博：@param content、  @param picPath、@throws WeiboException
private void share2Sinaweibo(String content, String picPath) throws WeiboException {
    Weibo weibo = Weibo.getInstance();
    weibo.share2weibo(this, weibo.getAccessToken().getToken(), weibo.getAccessToken()
            .getSecret(), content, picPath);
}
```

7.4 项目成果（见图 7-8 ~ 图 7-10）

图 7-8　程序主页面

图 7-9　当地游

图 7-10　详情页面

第8章 综合实训2 基于Android的手机定位项目

8.1 项目分析

无线通信行业的迅速发展，硬件技术和无线通信GPRS和CDMA网络的日趋成熟，手机的广泛应用，以及安卓手机的普及，使得手机定位我们的位置成为可能。用户无需其他的设备，只要拥有一部手机，就可以知道自己所在的位置。手机定位是指通过无线终端（手机）和无线网络配合，确定移动用户的实际位置信息（经/纬度坐标数据，包括三维数据），通过SMS、MMS、语音发给用户或以此为基础提供某种增值服务。定位服务又叫做移动位置服务（Location Based Service，LBS），是通过电信移动运营商的网络（如GSM网，CDMA网）获取移动终端用户的位置信息（经纬度坐标），在电子地图平台的支持下，为用户提供相应服务的一种增值业务。

手机定位按照运营商提供服务的方式可以分为两种：自有手机定位系统与公用定位服务。自有的定位系统主要是为企业和政府部门使用的定位系统，常用于对人员、事件、物品和车辆等的定位，这种定位方式广泛地用于公安执法、物流、长途车定位、紧急救援定位等；公用手机定位服务一般由移动运营商来提供，这种手机的定位有两种方式，一种是专线接入方式，另一种是短信告知方式。

8.2 项目设计

系统总体目标是通过此程序，用户可以用手机定位自己的位置，不用再担心到了一个陌生的城市迷路。其中，用户还能通过使用这个软件来查找特定的位置，规划去特定地方的路线，查看去过的地方，超出某个方位可以得到通知，将自己的信息分享给好友。

另外，由于手机软件的特殊性，存在一定的自然条件的限制，除了手机要保持网络畅通之外，该应用还应有以下几个功能：

（1）简单、友好的用户界面，保证用户直接上手便可以使用；
（2）操作简单、人性化、易用性高；

（3）支持 Internet 连接，保证定位准确；

（4）用户的历史记录使用 SQLite 进行存储，方便用户以后查看；

（5）向 PC 端开发靠拢，为以后 PC 与手机互联做准备，为将来人们能在 PC 端看到用户所在位置奠定基础。

"在哪儿"应用为给用户提供全方位的位置支持，主要分为"我在哪儿"、"电子地图"、"历史记录"、"周边搜索"、"线路规划"、"分享给好友"功能，如图 8-1 所示。

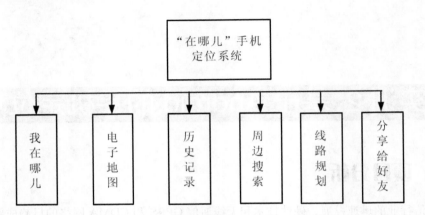

图 8-1 系统架构

图 8-1 中"我在哪儿"主要供用户对自己目前所在位置进行定位；"电子地图"能够为用户查询想去的地方的位置并在地图上显示出来；"历史记录"为用户记录曾经到达的地方；"周边搜索"供用户查询自己所在位置周围的餐厅、学校、ATM 机和公园；"线路规划"为用户规划如何从现在的位置到达目的地，"分享给好友"帮助用户通过短信将信息分享给好友。

系统详细架构如图 8-2 所示。

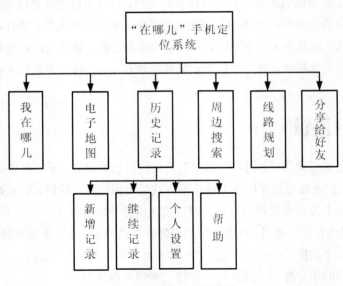

图 8-2 系统详细架构图

8.3 项目实施

开发架构关注软件开发环境下实际模块的组织。软件打包成小的程序块（程序库或子系统），它们可以由一位或几位开发人员来开发。子系统可以组织成分层结构，每个层为上一层提供良好定义的接口。

当前系统的文件结构如图 8-3 所示。

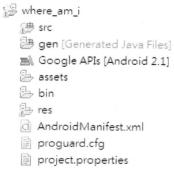

图 8-3 系统文件结构

其中，src 文件夹中有源程序包 com.where.am_i，其中包含本系统中所有用到的类，如图 8-4 所示，下面一一说明。

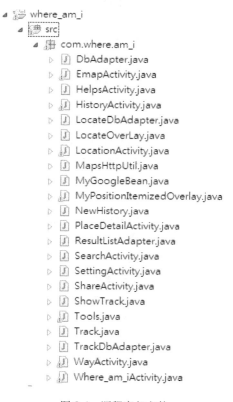

图 8-4 源程序包文件

- where_am_iActivity.java 是应用的主程序。
- LocationActivity.java 是实现"我在哪儿"功能的文件。
- EmapActivity.java 是实现"电子地图"功能的文件。
- HistoryActivity.java、NewHistory.java、SettingActivity.java、ShowTrack.java、Track.java、HelpsActivity.java 等是实现"历史记录"功能的文件。
- SearchActivity.java，MyGoogleBean.java，PlaceDetailActivity.java，ResultListAdapter.java，Tools.java 是实现"周边搜索"功能的文件。
- WayActivity 是实现"线路规划"功能的文件。
- DbAdapter.java, LocateDbAdapter.java, TrackDbAdapter.java 是实现数据库功能的文件。
- MyPositionItemizedOverlay.java，LocateOverLay.java 是实现在地图上进行标记功能的文件。

而 res 中为程序的页面和图片资源，res/layout 中为页面设置文件，如图 8-5 所示。

```
▲ res
   ▷  drawable-hdpi
   ▷  drawable-ldpi
   ▷  drawable-mdpi
   ▲  layout
         aboutuslayout.xml
         custom_dialog.xml
         helps.xml
         history_main.xml
         list_item.xml
         main.xml
         maplayout.xml
         new_history.xml
         opinionlayout.xml
         place_detail.xml
         searchmain.xml
         setting.xml
         sharelayout.xml
         show_track.xml
         track_row.xml
         waylayout.xml
         wherelayout.xml
   ▲  values
         array.xml
         helps.xml
         strings.xml
```

图 8-5　系统页面文件

- main.xml——首页。
- wherelayout.xml——"我在哪儿"页面。
- maplayout.xml——"电子地图"页面。
- history_main.xml——"历史记录"主页面。
- setting.xml——"历史记录"—"个人设置"页面。
- show_track.xml——展示历史记录页面。
- track_row.xml——展示历史记录边框。
- searchlayout.xml——"周边搜索"页面。
- place_detail.xml——店家详细信息页面。
- waylayout.xml——线路规划页面。
- sharelayout.xml——分享给好友页面。
- opinionlayout.xml——意见反馈页面。
- aboutuslayout.xml——关于我们页面。

8.3.1 我在哪儿

"我在哪儿"功能实现对手机位置的定位,用户可以通过单击首页的"我在哪儿"按钮,得到自己的位置,并且可以通过放大和缩小地图来查看周边的环境。该功能通过 LocationActivity.java 实现。通过 locate()方法定位。关键源代码如下。

public class LocationActivity **extends** MapActivity//从MapActivity基类派生子类
{
    public void onCreate(Bundle savedInstanceState) //子类重写基类onCreate方法
    {
        super.onCreate(savedInstanceState); //基类调用onCreate方法
        /*加载wherelayout。xml Layout*/
        setContentView(R.layout.*wherelayout*); //设置屏幕布局
        /*实例化相关视图组件*/
        linearLayout=(LinearLayout)findViewById(R.id.*zoomview*); //获得线性布局对象
        mapView=(MapView)findViewById(R.id.*mapview*); // 获得MapView对象
        mZoom=(ZoomControls)mapView.getZoomControls(); //获得ZoomControls对象
        linearLayout.addView(mZoom); //将控件对象添加到线性布局容器
        /*为当前所在点添加图层标示*/
        /*从MapView中获得MapController对象,调用locate()方法完成定位*/
        MapController controller=mapView.getController(); //获得MapController对象
        GeoPoint point=locate(controller); //获得GeoPoint对象信息
        mapOverlays=mapView.getOverlays();
        /*获得该图标对象Drawable*/

```
        drawable=this.getResources().getDrawable(R.drawable.androidmarker);
        itemizedOverlay=new MyPositionItemizedOverlay(drawable);
        OverlayItem overlayitem=new OverlayItem(point,"","");
        itemizedOverlay.addOverlay(overlayitem);
        mapOverlays.add(itemizedOverlay);
    }
    /*获得当前经纬度信息,通过MapController定位到该点*/
    private GeoPoint locate(MapController controller){
        locationManager=(LocationManager) getSystemService(Context.LOCATION_SERVICE);
        String provider=LocationManager.GPS_PROVIDER;
        Location location=locationManager.getLastKnownLocation(provider);
        double lat=0.0;
        double lng=0.0;
        if (location!=null){
            lat=location.getLatitude();
            lng=location.getLongitude();
        }else{
            lat=24.57;
            lng=121.13;
        }
        GeoPoint point=new GeoPoint((int)(lat*1E6),(int)(lng*1E6));
        controller.animateTo(point);
        return point;
}}}
```

8.3.2 电子地图

电子地图功能提供给用户查找位置,用户可以通过输入地址对该位置进行定位,并可以通过放大和缩小地图查看位置和周围环境。该功能通过 EmapActivity.java 文件实现。关键源代码如下。

```
public class EmapActivity extends MapActivity//从MapActivity基类派生子类EmapActivity
{/*获得当前经纬度信息,通过MapController定位到该点*/
    private void locate(GeoPoint point){//实现私有方法locate
        controller=mapView.getController();    //获得MapController对象
        controller.animateTo(point);
        drawable=this.getResources().getDrawable(R.drawable.androidmarker);
        mapOverlays=mapView.getOverlays();
        itemizedOverlay=new MyPositionItemizedOverlay(drawable);
```

```
        OverlayItem overlayitem= new OverlayItem(point,"","");
        itemizedOverlay.addOverlay(overlayitem);
        mapOverlays.add(itemizedOverlay);
    }
    public boolean onCreateOptionsMenu(Menu menu) {//重写菜单onCreateOptionsMenu方法
        menu.add(0,1,0,"显示输入查询地址对话框");
        return true;
    }
    public boolean onOptionsItemSelected(MenuItem item){//重写onOptionsItemSelected方法
        switch(item.getItemId()){
        case 1:
            customDialog();
            return true;
        }
        return false;
    }
    //根据地址获得GeoPoint对象
    private GeoPoint getGeoByAddress(String strSearchAddress)
    {
        GeoPoint gp=null;
        try{
            if(strSearchAddress!="")
            {Geocoder geoCoder=new Geocoder(EmapActivity.this, Locale.getDefault());
                List<Address>IstAddress=geoCoder.getFromLocationName(strSearchAddress,1);
                if(!IstAddress.isEmpty())
                {   Address adsLocation= IstAddress.get(0);
                    double geoLatitude=adsLocation.getLatitude()*1E6;
                    double geoLongitude=adsLocation.getLongitude()*1E6;
                    gp=new GeoPoint((int)geoLatitude,(int)geoLongitude);
                }}
        }
        catch(Exception e)//捕获异常
        {e.printStackTrace();}//显示异常信息
        return gp;
    }
    //自定义对话框
        private void customDialog(){
```

```java
        try{
            LayoutInflater mInflater01=LayoutInflater.from(EmapActivity.this);
            View myView=mInflater01.inflate(R.layout.custom_dialog,null);
            final EditText
addressEditText=(EditText)myView.findViewById(R.id.addressEditText);
            new AlertDialog.Builder(this)
                    .setView(myView).setTitle("请输入查询地址坐标")
                    .setPositiveButton("确定",
                new DialogInterface.OnClickListener(){
                    public void onClick(DialogInterface dialog, int whichButton){
                        GeoPoint point=getGeoByAddress(addressEditText.getText().toString());
                        locate(point);
                    }
            }).show();  //显示对话框
        }
        catch(Exception e)//捕获异常
        {e.printStackTrace();}//显示异常信息
}}
```

8.3.3 历史记录

历史记录功能可以记录用户的移动轨迹，并随时显示在 Google Map 上。当程序运行后，在历史记录主界面上创建一个新的记录，然后后台启动一个 service，定时读取 GPS 数据获得用户目前所在的位置信息，将其存入数据库中；用户可以选择以往的记录，将其轨迹显示在 Map 上，由此记录用户去过的位置。

该功能由 HistoryActivity.java、NewHistory.java、SettingActivity.java、ShowTrack.java、Track.java、HelpsActivity.java 等文件实现；而数据库存储由 DbAdapter.java、LocateDbAdapter.java、TrackDbAdapter.java 等文件实现。关键源代码如下。

（1）HistoryActivity.java 历史记录主页面

在 onCreate 方法中，使用了 render_tracks()方法从数据库中取出以往的记录，并更新到列表中，代码如下。

```java
public void onCreate(Bundle savedInstanceState) {//重写onCreate方法
    super.onCreate(savedInstanceState);   // 调用基类onCreate
    setContentView(R.layout.history_main);   //设置屏幕布局
    setTitle(R.string.app_title);   //设置标题
    mDbHelper = new TrackDbAdapter(this);   //创建TrackDbAdapter对象
    mDbHelper.open();   //打开数据库
```

render_tracks();
}

创建菜单以及菜单被选中后的相应方法，代码如下。

```
/*定义菜单需要的常量*/
    private static final int MENU_NEW=Menu.FIRST+1;
    private static final int MENU_CON=MENU_NEW+1;
    private static final int MENU_SETTING=MENU_CON+1;
    private static final int MENU_HELPS=MENU_SETTING+1;
    private static final int MENU_EXIT=MENU_HELPS+1;
/*初始化菜单*/
    public boolean onCreateOptionsMenu(Menu menu){//重写onCreateOptionsMenu方法
    super.onCreateOptionsMenu(menu);  //调用基类onCreateOptionsMenu方法
    /*如下添加几个菜单*/
    menu.add(0,MENU_NEW,0,R.string.menu_new).setIcon(
            R.drawable.new_track).setAlphabeticShortcut('N');
    menu.add(0,MENU_CON,0,R.string.menu_con).setIcon(
            R.drawable.con_track).setAlphabeticShortcut('C');
    menu.add(0,MENU_SETTING,0,R.string.menu_setting).setIcon(
            R.drawable.setting).setAlphabeticShortcut('S');
    menu.add(0,MENU_HELPS,0,R.string.menu_helps).setIcon(
            R.drawable.helps).setAlphabeticShortcut('H');
    menu.add(0,MENU_EXIT,0,R.string.menu_exit).setIcon(
            R.drawable.exit).setAlphabeticShortcut('E');
    return super.onCreateOptionsMenu(menu);
    }
/*当一个菜单被选中的时候调用*/
    public boolean onOptionsItemSelected(MenuItem item) {//重写onOptionsItemSelected方法
        Intent intent=new Intent();  //创建意图(Intent)对象
        switch (item.getItemId()){//处理菜单项
        case MENU_NEW:
                intent.setClass(HistoryActivity.this, NewHistory.class);  //新建菜单
                startActivity(intent);  //启动新的活动(Activity)
                return true;
        case MENU_CON:
                return true;
        case MENU_SETTING:
                intent.setClass(HistoryActivity.this, SettingActivity.class);  //设置功能处理
```

```
            startActivity(intent);  //启动新的活动(Activity)
            return true;
        case MENU_HELPS:
            intent.setClass(HistoryActivity.this, HelpsActivity.class);  //帮助功能
            startActivity(intent);  //启动新的活动(Activity)
        case MENU_EXIT://退出菜单
            finish();
        break;
        }
        return false;
}
```

当在列表中单击一个条目的时候调用 onListItemClick 方法,获取被单击条目的 ID、Name 和 Desc 属性,以绑定(Bundle)的方式传递给下一个活动,也就是实现单击以往跟踪记录条目的时候,打开一个新的活动,将路径展示在地图上,代码如下。

```
protected void onListItemClick(ListView l, View v, int position, long id) {//重写onListItemClick方法
        Log.d(TAG, "onListItemClick.");    //记录日志信息
        super.onListItemClick(l, v, position, id);  //调用基类onListItemClick方法
        Cursor c = mTrackCursor;  //处理游标
        c.moveToPosition(position);
        Intent i = new Intent(this, ShowTrack.class);  //新建意图对象
        i.putExtra(TrackDbAdapter.KEY_ROWID, id);  //绑定数据
        i.putExtra(TrackDbAdapter.NAME, c.getString(c
                .getColumnIndexOrThrow(TrackDbAdapter.NAME)));
        i.putExtra(TrackDbAdapter.DESC, c.getString(c
                .getColumnIndexOrThrow(TrackDbAdapter.DESC)));
        startActivity(i);  //切换到新的Activity
}
```

(2) NewHistory.java 新建记录

首先申明一些后面需要用到的变量,TAG 用来记录日志使用的标记,button_new 用来绑定模板中的"@+id/new_submit",field_new_name 和 field_new_desc 用来绑定模板中输入名字和描述信息的组件。然后再 onCreate 方法中设定其关联的 layout,实现代码如下所示。

```
public class NewHistory extends Activity{//从Activity基类派生子类NewHistory
    private static final String TAG="NewHistory";  //定义私有静态只读String常量
    private Button button_new;  //定义私有Button对象
    private EditText field_new_name;  //定义私有EditText对象
    private EditText field_new_desc;  //定义私有EditText对象
    private TrackDbAdapter mDbHelper;  //定义私有TrackDbAdapter对象
```

```java
public void onCreate(Bundle savedInstanceState) {//重写onCreate方法
    super.onCreate(savedInstanceState);  //调用基类onCreate方法
    setContentView(R.layout.new_history);  //设置屏幕布局
    setTitle(R.string.menu_new);  //设置屏幕标题
    findViews();
    setListensers();
    mDbHelper = new TrackDbAdapter(this);
    mDbHelper.open();
}
```

其中的findViews()用来查找模板中需要用到的控件,其实现代码如下。

```java
private void findViews() {
    Log.d(TAG, "find Views");
    field_new_name = (EditText) findViewById(R.id.new_name);
    field_new_desc = (EditText) findViewById(R.id.new_desc);
    button_new = (Button) findViewById(R.id.new_submit);
}
```

onCreate 方法中的 setListensers 方法用来设定监听器,这里直接调用 button 的 setOnClickListener()方法设置其单击的相应监听器为 new_track,代码如下。

```java
// Listen for button clicks
private void setListensers() {
    Log.d(TAG, "set Listensers");
    button_new.setOnClickListener(new_track);
}
```

上面设定其监听器为 new_track,其监听器的实现逻辑代码如下。

```java
private Button.OnClickListener new_track = new Button.OnClickListener() {
    public void onClick(View v) {
        Log.d(TAG, "onClick new_track..");
        try {
            String name = (field_new_name.getText().toString());
            String desc = (field_new_desc.getText()
                    .toString());
            if (name.equals("")) {
                Toast.makeText(NewHistory.this,
                        getString(R.string.new_name_null),
                        Toast.LENGTH_SHORT).show();
            } else {
                //调用存储接口保存到数据库并启动service
```

```
                    Long row_id = mDbHelper.createTrack(name, desc);
                    Log.d(TAG, "row_id="+row_id);
                    Intent intent = new Intent();
                    intent.setClass(NewHistory.this, ShowTrack.class);
                    intent.putExtra(TrackDbAdapter.KEY_ROWID, row_id);
                    intent.putExtra(TrackDbAdapter.NAME, name);
                    intent.putExtra(TrackDbAdapter.DESC, desc);
                    startActivity(intent);
                }
            } catch (Exception err) {
                Log.e(TAG, "error: " + err.toString());
                Toast.makeText(NewHistory.this, getString(R.string.new_fail),
                    Toast.LENGTH_SHORT).show();
            }}};
```

（3）SettingActivity.java 个人设置

先声明一些需要使用的变量，然后使用 setContentView()设定其对应的布局文件为 setting.xml，使用 setTitle()设定其标题，进而调用 findViews()找到我们需要操作的组件，并调用 setListensers()给按钮设定单击监听器，最后调用 restorePrefs()将默认值或用户以前的选择值显示出来，代码如下：

```
public class SettingActivity extends Activity {//从Activity基类派生子类SettingActivity
    private static final String TAG = "Setting";
    //定义菜单需要的常量
    private static final int MENU_MAIN = Menu.FIRST + 1;
    private static final int MENU_NEW = MENU_MAIN + 1;
    private static final int MENU_BACK = MENU_NEW + 1;;
    // 保存个性化设置
    public static final String SETTING_INFOS = "SETTING_Infos";
    public static final String SETTING_GPS = "SETTING_Gps";
    public static final String SETTING_MAP = "SETTING_Map";
    public static final String SETTING_GPS_POSITON = "SETTING_Gps_p";
    public static final String SETTING_MAP_POSITON = "SETTING_Map_p";
    private Button button_setting_submit;
    private Spinner field_setting_gps;
    private Spinner field_setting_map_level;
    public void onCreate(Bundle savedInstanceState) {//重写onCreate方法
        super.onCreate(savedInstanceState);   //调用基类onCreate方法
        setContentView(R.layout.setting);   //设置屏幕布局
setTitle(R.string.menu_setting);   //设置屏幕标题
```

```
        findViews();    //找到所需要操作的组件
        setListensers();  //给按钮设定单击监听器
        restorePrefs();  //将默认值或用户以前的选择值显示出来
    }
    private void findViews() {//找到所需要操作的组件
        Log.d(TAG, "find Views");
        button_setting_submit = (Button) findViewById(R.id.setting_submit);
        field_setting_gps = (Spinner) findViewById(R.id.setting_gps);
        ArrayAdapter<CharSequence> adapter = ArrayAdapter.createFromResource(
                this, R.array.gps, android.R.layout.simple_spinner_item);
        adapter.setDropDownViewResource(android.R.layout.simple_spinner_dropdown_item);
        field_setting_gps.setAdapter(adapter);
        field_setting_map_level = (Spinner) findViewById(R.id.setting_map_level);
        ArrayAdapter<CharSequence> adapter2 = ArrayAdapter.createFromResource(
                this, R.array.map, android.R.layout.simple_spinner_item);
        adapter2.setDropDownViewResource(android.R.layout.simple_spinner_dropdown_item);
        field_setting_map_level.setAdapter(adapter2);
    }
```

上述代码在视图中找到一个 Button 和两个 Spinner 组件，然后采取调用 Spinner 的 setDropDownViewResource 方法设定其下拉框内容，这里两个 Spinner 分别使用 R.array.gps 和 R.array.map 数据源，这两个数据源是预先定义好，放在 value/array.xml 中的，其实现代码如下所示。

```xml
<resources>
    <!-- Used in View/setting.java -->
    <string-array name="gps">
        <item>5</item>
        <item>10</item>
        <item>15</item>
        <item>30</item>
        <item>60</item>
        <item>120</item>
    </string-array>
    <string-array name="map">
        <item>5</item>
        <item>6</item>
        <item>7</item>
        <item>8</item>
        <item>9</item>
```

```xml
            <item>10</item>
            <item>11</item>
            <item>12</item>
            <item>13</item>
            <item>14</item>
            <item>15</item>
            <item>16</item>
        </string-array>
</resources>
```

onCreate 方法中调用了 setListeners 设定监听器 setting_submit。监听器监听按钮被单击的事件实现如下：首先创建一个 Button.OnClickListener()对象 setting_submit，进而实现其 onClick()方法，在 onclick()中，使用 getSelectedItem 获得选择的条目，进行一些基本的判断，如果信息不完整，则使用 Toast 显示预先定义的提示消息，如果没有问题，则使用 storePrefs()进行存储。这里就是使用 SharedPreferences 来存储配置信息的，除了保存选择的配置，还保存了其配置在列表中的位置，以方便后面使用。

```java
//给按钮设定单击监听器
    private void setListensers() {
        Log.d(TAG, "set Listensers");
        button_setting_submit.setOnClickListener(setting_submit);
    }
    private Button.OnClickListener setting_submit = new Button.OnClickListener() {
        public void onClick(View v) {
            Log.d(TAG, "onClick new_track..");
            try {
                String gps = (field_setting_gps.getSelectedItem().toString());
                String map = (field_setting_map_level.getSelectedItem()
                        .toString());
                if (gps.equals("") || map.equals("")) {
                    Toast.makeText(SettingActivity.this,
                            getString(R.string.setting_null),
                            Toast.LENGTH_SHORT).show();
                } else {
                    storePrefs();  //保存设定
                    Toast.makeText(SettingActivity.this,
                            getString(R.string.setting_ok),
                            Toast.LENGTH_SHORT).show();
                    //跳转到主界面
```

```java
                    Intent intent = new Intent();
                    intent.setClass(SettingActivity.this, HistoryActivity.class);
                    startActivity(intent);
                }
            } catch (Exception err) {
                Log.e(TAG, "error: " + err.toString());
                Toast.makeText(SettingActivity.this, getString(R.string.setting_fail),
                        Toast.LENGTH_SHORT).show();
            }}};
//保存个人设置
    private void storePrefs() {
        Log.d(TAG, "storePrefs setting infos");
        SharedPreferences settings = getSharedPreferences(SETTING_INFOS, 0);
        settings.edit()
            .putString(SETTING_GPS, field_setting_gps.getSelectedItem().toString())
            .putString(SETTING_MAP, field_setting_map_level.getSelectedItem().toString())
            .putInt(SETTING_GPS_POSITON, field_setting_gps.getSelectedItemPosition())
            .putInt(SETTING_MAP_POSITON, field_setting_map_level.getSelectedItemPosition())
            .commit();
    }
```

在 onCreate 中还有一个过程 restorePrefs，它负责将以前的设定显示在下拉选择的默认位置上，其实现代码如下。

```java
// Restore preferences
    private void restorePrefs() {
        SharedPreferences settings = getSharedPreferences(SETTING_INFOS, 0);
        int setting_gps_p = settings.getInt(SETTING_GPS_POSITON, 0);
        int setting_map_level_p = settings.getInt(SETTING_MAP_POSITON, 0);
        Log.d(TAG, "restorePrefs: setting_gps= "+ setting_gps_p + ",setting_map_level=" + setting_map_level_p);
        //如果值存在，就把其设定给指定的组件
        if (setting_gps_p != 0 && setting_map_level_p != 0) {
            field_setting_gps.setSelection(setting_gps_p);
            field_setting_map_level.setSelection(setting_map_level_p);
            button_setting_submit.requestFocus();
        }else if(setting_gps_p != 0 ){
            field_setting_gps.setSelection(setting_gps_p);
            field_setting_map_level.requestFocus();
```

```java
        }else if(setting_map_level_p != 0){
            field_setting_map_level.setSelection(setting_map_level_p);
            field_setting_gps.requestFocus();
        }else{
            field_setting_gps.requestFocus();
        }}
```

实现创建菜单和选择菜单,在程序退出的时候顺便保存设置,代码如下。

```java
// 初始化菜单
    public boolean onCreateOptionsMenu(Menu menu) {
        super.onCreateOptionsMenu(menu);
        menu.add(0, MENU_MAIN, 0, R.string.menu_main).setIcon(
                R.drawable.icon).setAlphabeticShortcut('M');
        menu.add(0, MENU_NEW, 0, R.string.menu_new).setIcon(
                R.drawable.new_track).setAlphabeticShortcut('N');
        menu.add(0, MENU_BACK, 0, R.string.menu_back).setIcon(
                R.drawable.back).setAlphabeticShortcut('E');
        return true;
    }
// 当一个菜单被选中的时候调用
    public boolean onOptionsItemSelected(MenuItem item) {
        Intent intent = new Intent();
        switch (item.getItemId()) {
        case MENU_NEW:
            intent.setClass(SettingActivity.this, NewHistory.class);
            startActivity(intent);
            return true;
        case MENU_MAIN:
            intent.setClass(SettingActivity.this, HistoryActivity.class);
            startActivity(intent);
            return true;
        case MENU_BACK:
            finish();
            break;
        }
        return true;}}
    protected void onStop(){
        super.onStop();
```

```
        Log.d(TAG, "save setting infos");
        storePrefs();
}
```

（4）ShowTrack.java 地图展示界面

首先申明一些需要用到的变量，实现 onCreate()方法和 findViews()方法，代码如下。

```
public class ShowTrack extends MapActivity {//从基类MapActivity派生子类ShowTrack
    // 定义菜单需要的常量
    private static final int MENU_NEW = Menu.FIRST + 1;
    private static final int MENU_CON = MENU_NEW + 1;
    private static final int MENU_DEL = MENU_CON + 1;
    private static final int MENU_MAIN = MENU_DEL + 1;
    private TrackDbAdapter mDbHelper;
    private LocateDbAdapter mlcDbHelper;
    private static final String TAG = "ShowTrack";
    private static MapView mMapView;
    private MapController mc;
    protected MyLocationOverlay mOverlayController;
    private Button mZin;
    private Button mZout;
    private Button mPanN;
    private Button mPanE;
    private Button mPanW;
    private Button mPanS;
    private Button mGps;
    private Button mSat;
    private Button mTraffic;
    private Button mStreetview;
    private String mDefCaption = "";
    private GeoPoint mDefPoint;
    private LocationManager lm;
    private LocationListener locationListener;
    private int track_id;
    private Long rowId;
    public void onCreate(Bundle icicle) {//重写onCreate方法
        super.onCreate(icicle); //调用基类onCreate方法
        setContentView(R.layout.show_track); //设置屏幕布局
        findViews();
```

Android 应用程序设计

```
        centerOnGPSPosition();
        revArgs();
        paintLocates();
        startTrackService();
}
```

地图模式切换的实现代码如下。

```java
public void toggleSatellite() {
        mMapView.setSatellite(true);
        mMapView.setStreetView(false);
        mMapView.setTraffic(false);
}
public void toggleTraffic() {
        mMapView.setTraffic(true);
        mMapView.setSatellite(false);
        mMapView.setStreetView(false);
}
public void toggleStreetView() {
        mMapView.setStreetView(true);
        mMapView.setSatellite(false);
        mMapView.setTraffic(false);
}
```

要将地图定位到当前的 GPS 指定的位置，使用 LocationManager 得到当前的 GPS 位置信息，然后将地图定位过去，并使用基于 Overlay 的 MyOverlay 将其显示在地图上。Android 中的 Overlay 是个抽象类，必须重载并实现其 draw 方法。在这儿首先使用 toPixels 将 GPS 获得的 Geopoint 点转换为手机屏幕上的点坐标 myStreetCoords，然后使用 Android 的 draw 系列画出需要的标识，代码如下。

```java
private void centerOnGPSPosition() {
        Log.d(TAG, "centerOnGPSPosition");
        String provider = "gps";
        LocationManager lm = (LocationManager)
                getSystemService(Context.LOCATION_SERVICE);
        Location loc = lm.getLastKnownLocation(provider);
        loc = lm.getLastKnownLocation(provider);
        mDefPoint = new GeoPoint((int) (loc.getLatitude() * 1000000),
                (int) (loc.getLongitude() * 1000000));
        mDefCaption = "I'm Here.";
        mc.animateTo(mDefPoint);
        mc.setCenter(mDefPoint);
```

```java
        MyOverlay mo = new MyOverlay();
        mo.onTap(mDefPoint, mMapView);
        mMapView.getOverlays().add(mo);
    }
    protected class MyOverlay extends Overlay {
        public void draw(Canvas canvas, MapView mv, boolean shadow) {
            Log.d(TAG, "MyOverlay::darw..mDefCaption=" + mDefCaption);
            super.draw(canvas, mv, shadow);
            if (mDefCaption.length() == 0) {
                return;}
            Paint p = new Paint();
            int[] scoords = new int[2];
            int sz = 5;
            Point myScreenCoords = new Point();
            mMapView.getProjection().toPixels(mDefPoint, myScreenCoords);
            scoords[0] = myScreenCoords.x;
            scoords[1] = myScreenCoords.y;
            p.setTextSize(14);
            p.setAntiAlias(true);
            int sw = (int) (p.measureText(mDefCaption) + 0.5f);
            int sh = 25;
            int sx = scoords[0] - sw / 2 - 5;
            int sy = scoords[1] - sh - sz - 2;
            RectF rec = new RectF(sx, sy, sx + sw + 10, sy + sh);
            p.setStyle(Style.FILL);
            p.setARGB(128, 255, 0, 0);
            canvas.drawRoundRect(rec, 5, 5, p);
            p.setStyle(Style.STROKE);
            p.setARGB(255, 255, 255, 255);
            canvas.drawRoundRect(rec, 5, 5, p);
            canvas.drawText(mDefCaption, sx + 5, sy + sh - 8, p);
            p.setStyle(Style.FILL);
            p.setARGB(88, 255, 0, 0);
            p.setStrokeWidth(1);
            RectF spot = new RectF(scoords[0] - sz, scoords[1] + sz, scoords[0]
                    + sz, scoords[1] - sz);
            canvas.drawOval(spot, p);
```

```
            p.setARGB(255, 255, 0, 0);
            p.setStyle(Style.STROKE);
            canvas.drawCircle(scoords[0], scoords[1], sz, p);
        }
    }
```
最后需要实现的是前面要用的 MyLocationListener，代码如下。
```
protected class MyLocationListener implements LocationListener {//实现LocationListener接口
    public void onLocationChanged(Location loc) {//重写onLocationChanged
        Log.d(TAG, "MyLocationListener::onLocationChanged..");
        if (loc != null) {
            Toast.makeText(
                    getBaseContext(),
                    "Location changed : Lat: " + loc.getLatitude()
                            + " Lng: " + loc.getLongitude(),
                    Toast.LENGTH_SHORT).show();
            mDefPoint = new GeoPoint((int) (loc.getLatitude() * 1000000),
                    (int) (loc.getLongitude() * 1000000));
            mc.animateTo(mDefPoint);
            mc.setCenter(mDefPoint);
            mDefCaption = "Lat: " + loc.getLatitude() + ",Lng: "
                    + loc.getLongitude();
            MyOverlay mo = new MyOverlay();
            mo.onTap(mDefPoint, mMapView);
            mMapView.getOverlays().add(mo);
        }
    }
    public void onProviderDisabled(String provider) {//重写onProviderDisabled方法
        Toast.makeText(
                getBaseContext(),
                "ProviderDisabled.",
                Toast.LENGTH_SHORT).show();        }
    public void onProviderEnabled(String provider) {
        Toast.makeText(
                getBaseContext(),
                "ProviderEnabled,provider:"+provider,
                Toast.LENGTH_SHORT).show();
    public void onStatusChanged(String provider, int status, Bundle extras) {
```

        }
    }
（5）DbAdapter.java

在这个类中完成两个表的创建和升级等操作，重新定义 SQLiteOpenHelper 的 onCreate 和 onUpdate 方法，在这两种方法中编写创建和升级数据库的脚本，代码如下。

```java
public class DbAdapter {//实现数据库类DbAdapter
    private static final String TAG = "DbAdapter";
    private static final String DATABASE_NAME = "iTracks.db";
    private static final int DATABASE_VERSION = 1;
    public class DatabaseHelper extends SQLiteOpenHelper {//从SQLiteOpenHelper基类派生子类
        public DatabaseHelper(Context context) {//构造器
            super(context, DATABASE_NAME, null, DATABASE_VERSION); //调用基类构造器
        }
        public void onCreate(SQLiteDatabase db) {//重写onCreate方法
            String tracks_sql = "CREATE TABLE " + TrackDbAdapter.TABLE_NAME + " ("
                + TrackDbAdapter.ID     + " INTEGER primary key autoincrement, "
                + TrackDbAdapter.NAME+ " text not null, "
                + TrackDbAdapter.DESC + " text ,"
                + TrackDbAdapter.DIST + " LONG ,"
                + TrackDbAdapter.TRACKEDTIME + " LONG ,"
                + TrackDbAdapter.LOCATE_COUNT + " INTEGER, "
                + TrackDbAdapter.CREATED + " text, "
                + TrackDbAdapter.AVGSPEED + " LONG, "
                + TrackDbAdapter.MAXSPEED + " LONG ,"
                + TrackDbAdapter.UPDATED + " text "
                + ");";
            Log.i(TAG, tracks_sql);
            db.execSQL(tracks_sql);
            String locats_sql = "CREATE TABLE " + LocateDbAdapter.TABLE_NAME + " ("
                + LocateDbAdapter.ID     + " INTEGER primary key autoincrement, "
                + LocateDbAdapter.TRACKID + " INTEGER not null, "
                + LocateDbAdapter.LON + " DOUBLE ,"
                + LocateDbAdapter.LAT + " DOUBLE ,"
                + LocateDbAdapter.ALT + " DOUBLE ,"
                + LocateDbAdapter.CREATED + " text "
                + ");";
            Log.i(TAG, locats_sql);
```

```
            db.execSQL(locats_sql);
    }
        public void onUpgrade(SQLiteDatabase db, int oldVersion, int newVersion) {
            db.execSQL("DROP TABLE IF EXISTS " + LocateDbAdapter.TABLE_NAME + ";");
            db.execSQL("DROP TABLE IF EXISTS " + TrackDbAdapter.TABLE_NAME + ";");
            onCreate(db);
    }}}
```

（6）TrackDbAdapter.java

代码一开始声明一些常量，然后按照操作需要定义了集中操作，如 createTrack 用来创建新的跟踪记录，updateTrack 用来更新已有记录等，代码如下。

```
public class TrackDbAdapter extends DbAdapter{//从DbAdapter基类派生子类TrackDbAdapter
    private static final String TAG = "TrackDbAdapter";
    public static final String TABLE_NAME = "tracks";
    public static final String ID = "_id";
    public static final String KEY_ROWID = "_id";
    public static final String NAME = "name";
    public static final String DESC = "desc";
    public static final String DIST = "distance";
    public static final String TRACKEDTIME = "tracked_time";
    public static final String LOCATE_COUNT = "locats_count";
    public static final String CREATED = "created_at";
    public static final String UPDATED = "updated_at";
    public static final String AVGSPEED = "avg_speed";
    public static final String MAXSPEED = "max_speed";
    private DatabaseHelper mDbHelper;
    private SQLiteDatabase mDb;
    private final Context mCtx;
    public TrackDbAdapter(Context ctx) {//构造器
        this.mCtx = ctx;
    }
    public TrackDbAdapter open() throws SQLException {//打开数据库
        mDbHelper = new DatabaseHelper(mCtx);
        mDb = mDbHelper.getWritableDatabase();
        return this;
    }
    public void close() {//关闭数据库
        mDbHelper.close();
```

```java
    }
    public Cursor getTrack(long rowId) throws SQLException {
        Cursor mCursor =
        mDb.query(true, TABLE_NAME, new String[] { KEY_ROWID, NAME,
                DESC, CREATED }, KEY_ROWID + "=" + rowId, null, null,
                null, null, null);
        if (mCursor != null) {
            mCursor.moveToFirst();
        }
        return mCursor;
    }
    public long createTrack(String name, String desc) {
        Log.d(TAG, "createTrack.");
        ContentValues initialValues = new ContentValues();
        initialValues.put(NAME, name);
        initialValues.put(DESC, desc);
        Calendar calendar = Calendar.getInstance();
        String created = calendar.get(Calendar.YEAR) + "-" +calendar.get(Calendar.MONTH) + "-" + calendar.get(Calendar.DAY_OF_MONTH) + " "
                + calendar.get(Calendar.HOUR_OF_DAY) + ":"
                + calendar.get(Calendar.MINUTE) + ":"   + calendar.get(Calendar.SECOND);
        initialValues.put(CREATED, created);
        initialValues.put(UPDATED, created);
        return mDb.insert(TABLE_NAME, null, initialValues);
    }
    public boolean deleteTrack(long rowId) {
        return mDb.delete(TABLE_NAME, KEY_ROWID + "=" + rowId, null) > 0;
    }
    public Cursor getAllTracks() {
        return mDb.query(TABLE_NAME, new String[] { ID, NAME,
                DESC, CREATED }, null, null, null, null, "updated_at desc");
    }
    public boolean updateTrack(long rowId, String name, String desc) {
        ContentValues args = new ContentValues();
        args.put(NAME, name);
        args.put(DESC, desc);
        Calendar calendar = Calendar.getInstance();
```

Android 应用程序设计

```java
            String updated = calendar.get(Calendar.YEAR) + "-" +calendar.get(Calendar.MONTH) + "-" +
calendar.get(Calendar.DAY_OF_MONTH) + " "
            + calendar.get(Calendar.HOUR_OF_DAY) + ":"
            + calendar.get(Calendar.MINUTE) + ":"  + calendar.get(Calendar.SECOND);
        args.put(UPDATED, updated);
        return mDb.update(TABLE_NAME, args, KEY_ROWID + "=" + rowId, null) > 0;
    }
}
```

（7）LocateDbAdapter.java

和 TrackDbAdapter 一样，先定义一些常量，然后按照操作的需要定义几种操作，代码如下。

```java
public class LocateDbAdapter extends DbAdapter {
    private static final String TAG = "LocateDbAdapter";
    public static final String TABLE_NAME = "locates";
    public static final String ID = "_id";
    public static final String TRACKID = "track_id";
    public static final String LON = "longitude";
    public static final String LAT = "latitude";
    public static final String ALT = "altitude";
    public static final String CREATED = "created_at";
    private DatabaseHelper mDbHelper;
    private SQLiteDatabase mDb;
    private final Context mCtx;
    public LocateDbAdapter(Context ctx) {
        this.mCtx = ctx;
    }
    public LocateDbAdapter open() throws SQLException {
        mDbHelper = new DatabaseHelper(mCtx);
        mDb = mDbHelper.getWritableDatabase();
        return this;
    }
    public void close() {
        mDbHelper.close();
    }
    public Cursor getLocate(long rowId) throws SQLException {
        Cursor mCursor =
            mDb.query(true, TABLE_NAME, new String[] { ID, LON,
                LAT, ALT, CREATED }, ID + "=" + rowId, null, null,
```

220

```
            null, null, null);
        if (mCursor != null) {
            mCursor.moveToFirst();
        }
        return mCursor;
    }
    public long createLocate(int track_id, Double longitude, Double latitude, Double altitude) {
        Log.d(TAG, "createLocate.");
        ContentValues initialValues = new ContentValues();
        initialValues.put(TRACKID, track_id);
        initialValues.put(LON, longitude);
        initialValues.put(LAT, latitude);
        initialValues.put(ALT, altitude);
        Calendar calendar = Calendar.getInstance();
        String created = calendar.get(Calendar.YEAR) + "-" +calendar.get(Calendar.MONTH) + "-" + calendar.get(Calendar.DAY_OF_MONTH) + " "
                + calendar.get(Calendar.HOUR_OF_DAY) + ":"
                + calendar.get(Calendar.MINUTE) + ":"   + calendar.get(Calendar.SECOND);
        initialValues.put(CREATED, created);
        return mDb.insert(TABLE_NAME, null, initialValues);
    }
    public boolean deleteLocate(long rowId) {
        return mDb.delete(TABLE_NAME, ID + "=" + rowId, null) > 0;
    }
    public Cursor getTrackAllLocates(int trackId) {
        return mDb.query(TABLE_NAME, new String[] { ID,TRACKID, LON,
                LAT, ALT,CREATED }, "track_id=?", new String[] {String.valueOf(trackId)}, null, null, "created_at asc");}}
```

（8）Track.java（实现 Service）

因为想要的 GPS 效果是跟踪开始后，切换界面等操作不影响 GPS 跟踪，也就是希望可以在后台不间断地跟踪、记录，所以需要用到 Service。让 Track 继承 Android 的 Service 类，然后在其 onStart 中连接数据库，接收参数并设定监听器，并使用 MyLocationListener，使得在 onLocationChanged 的时候，调用前面数据存储部分已经实现的 mlcDbHelper.creatLocate(track_id,loc.getLongitude(),loc.getLatitude(),loc.getAlttude())，将位置信息和接收到的参数写入数据库中，再加上与之相对应的启动和停止 Service 的方法，只要启动了 Service，其就会在后台一直运行，当位置信息发生变化时，调用数据库存储接口存入数据库，具体代码如下。

```
public class Track extends Service {//从基类Service派生子类Track
```

```java
        private static final String TAG = "Track";
        private LocationManager lm;
        private LocationListener locationListener;
        static LocateDbAdapter mlcDbHelper = null;
        private int track_id;
        public IBinder onBind(Intent arg0) {
            Log.d(TAG, "onBind.");
            return null;
        }
        public void onStart(Intent intent, int startId) {
            Log.d(TAG, "onStart.");
            super.onStart(intent, startId);
            startDb();
            Bundle extras = intent.getExtras();
            if (extras != null) {
                track_id = extras.getInt(LocateDbAdapter.TRACKID);
            }
            Log.d(TAG, "track_id =" + track_id);
            lm = (LocationManager) getSystemService(Context.LOCATION_SERVICE);
            locationListener = new MyLocationListener();
            lm.requestLocationUpdates(LocationManager.GPS_PROVIDER, 0, 0,
                    locationListener);
        }
        private void startDb() {
            if(mlcDbHelper == null){
                mlcDbHelper = new LocateDbAdapter(this);
                mlcDbHelper.open();
            }
        }
        private void stopDb() {
            if(mlcDbHelper != null){
                mlcDbHelper.close();
            }}
        public static LocateDbAdapter getDbHelp(){
            return mlcDbHelper;
        }
        public void onDestroy() {
```

```
        Log.d(TAG, "onDestroy.");
        super.onDestroy();
        stopDb(); }
    protected class MyLocationListener implements LocationListener {
        public void onLocationChanged(Location loc) {
            Log.d(TAG, "MyLocationListener::onLocationChanged..");
            if (loc != null) {
                if(mlcDbHelper == null){
                    mlcDbHelper.open();
                }
                mlcDbHelper.createLocate(track_id,   loc.getLongitude(),loc.getLatitude(),loc.getAltitude());
            }}
        public void onProviderDisabled(String provider) {
            Toast.makeText(
                    getBaseContext(),
                    "ProviderDisabled.",
                    Toast.LENGTH_SHORT).show();          }
        public void onProviderEnabled(String provider) {
            Toast.makeText(
                    getBaseContext(),
                    "ProviderEnabled,provider:"+provider,
                    Toast.LENGTH_SHORT).show();          }
        public void onStatusChanged(String provider, int status, Bundle extras) {
            // TODO ：重写onStatusChanged方法
        }}}
```

8.3.4 周边搜索

周边搜索功能提供给用户这样一个功能，用户可以通过该功能来搜索所在位置周围的餐厅、ATM 机、学校和公园。该功能通过 SearchActivity.java 文件实现。

在 SearchActivity.java 中先定义四个方法，从服务器中分别获得 2000m 内的餐厅、ATM 机、学校和公园的信息。关键源代码如下。

```
public class SearchActivity extends Activity
    String resultString;
    ListView resultListView;
    private ResultListAdapter resultListAdapter;  //从基类Activity派生子类SearchActivity
```

Android 应用程序设计

```java
ProgressDialog progressDialog;
public void onCreate(Bundle savedInstanceState) {
    super.onCreate(savedInstanceState);
    setContentView(R.layout.searchmain);
    progressDialog = new ProgressDialog(SearchActivity.this);
    progressDialog.setMessage("Loading...");
    resultListView = (ListView) findViewById(R.id.resultList);
    adapterListener(resultListView);
}
public void toUniversity(View view) {
    progressDialog.show();
    new GetMessageFromServer().execute("2000,university");
}
public void toPark(View view) {
    progressDialog.show();
    new GetMessageFromServer().execute("2000,park");
}
public void toFood(View view) {
    progressDialog.show();
    new GetMessageFromServer().execute("2000,food");
}
public void toATM(View view) {
    progressDialog.show();
    new GetMessageFromServer().execute("2000,atm");
}
public void adapterListener(ListView listView) {
    listView.setOnItemClickListener(new OnItemClickListener() {
        public void onItemClick(AdapterView<?> parent, View view, int position, long id) {
            try {
                String refranceString =
                    Tools.formatJsonTOBean(resultString).get(position).getReference();
                Intent intent = new Intent(SearchActivity.this,PlaceDetailActivity.class);
                intent.putExtra(Tools.PLACE_REFRANCE, refranceString);
                startActivity(intent);
            } catch (JSONException e) {
                e.printStackTrace();
            }
        }});}
```

```java
private class GetMessageFromServer extends AsyncTask<String, Void, String> {
    protected String doInBackground(String... parameters) {
        try {
            resultString = MapsHttpUtil.getGetRoundPlace(Tools.getLocation(getApplicationContext()),
                parameters[0].split(",")[0],parameters[0].split(",")[1] );
        } catch (IOException e) {
            e.printStackTrace();
        }
        return resultString;
    }
    protected void onPostExecute(String result) {
        try {
            resultListAdapter = new ResultListAdapter(Tools.formatJsonTOBean(resultString), getApplicationContext());
            resultListView.setAdapter(resultListAdapter);
            progressDialog.dismiss();
        } catch (JSONException e) {
            e.printStackTrace(); }
        super.onPostExecute(result);}
}}
```

在 Tools.java 中获得用户当前的位置，以定位获得周围的店家，代码如下。

```java
public class Tools {
    public static final String PLACE_REFRANCE = "refranceString";
    public static String getLocation(Context context){
        LocationManager locationManager = (LocationManager)context.getSystemService(Context.LOCATION_SERVICE);
        Location location = locationManager.getLastKnownLocation(LocationManager.NETWORK_PROVIDER);
        if(location == null){
            return "24.970463,121.266947";
        }
        double longitude = location.getLongitude();
        double latitude = location.getLatitude();
        Log.i("tag", "location is " + longitude + "," + latitude);
        return latitude+","+longitude;          }
```

Android 应用程序设计

```java
public static List<MyGoogleBean> formatJsonTOBean(String resultString) throws JSONException{
    List<MyGoogleBean> myGoogleBeans = new ArrayList<MyGoogleBean>();
    JSONObject clientResonseJsonObjetc = new JSONObject(resultString);
    JSONArray jsonArray = clientResonseJsonObjetc.getJSONArray("results");
    for (int i = 0; i < jsonArray.length(); i++) {
        JSONObject jsonObject = jsonArray.getJSONObject(i);
        String name = jsonObject.getString("name");
        String address = jsonObject.getString("vicinity");
        String reference = jsonObject.getString("reference");
        myGoogleBeans.add(new MyGoogleBean(name, address,reference));
    }
    return myGoogleBeans;}  }
```

在 ResultListAdapter.java 中获得相应的结果的列表，并显示出来，代码如下。

```java
public class ResultListAdapter extends BaseAdapter{
    private List<MyGoogleBean> myGoogleBeans;
    private Context context;
    public ResultListAdapter(List<MyGoogleBean> myGoogleBeans, Context context) {
        super();
        this.myGoogleBeans = myGoogleBeans;
        this.context = context;
    }
    public int getCount() {
        return myGoogleBeans.size();}
    public Object getItem(int position) {
        return myGoogleBeans.get(position);}
    public long getItemId(int position) {
        return 0;}
    public View getView(int position, View convertView, ViewGroup parent) {
        if(convertView == null){
            convertView = (View)LayoutInflater.from(context).inflate(R.layout.list_item,null);
        }
        TextView nameTextView = (TextView)convertView.findViewById(R.id.nameString);
        TextView addTextView = (TextView)convertView.findViewById(R.id.addressString);
        nameTextView.setText(myGoogleBeans.get(position).getName());
        addTextView.setText(myGoogleBeans.get(position).getAddress());
        return convertView;
```

}}

通过 MyGoogleBean.Java 从 API 中获得相应的信息,如被选中店家的名称、地址、电话,代码如下。

```java
public class MyGoogleBean {
private String name;
    private String address;
    private String reference;
    public String getReference() {
        return reference;
    }
    public void setReference(String reference) {
        this.reference = reference;
    }
    public String getName() {
        return name;
    }
    public void setName(String name) {
        this.name = name;
    }
    public String getAddress() {
        return address;
    }
    public void setAddress(String address) {
        this.address = address;
    }
    public MyGoogleBean(String name, String address, String reference) {
        super();
        this.name = name;
        this.address = address;
        this.reference = reference;
    }}
```

在 PlaceDetailActivity.java 中获得被选中的店家的详细信息,如果用户还想知道更多,可以通过网络搜寻得到,代码如下。

```java
public class PlaceDetailActivity extends Activity{
    String refranceString;
    TextView placeName, placeAddress, phoneNumber;
    String mapURL;
    public void onCreate(Bundle savedInstanceState) {
```

```
        super.onCreate(savedInstanceState);
        setContentView(R.layout.place_detail);
        placeName = (TextView) findViewById(R.id.detailNameString);
        placeAddress = (TextView) findViewById(R.id.detailAddressString);
        phoneNumber = (TextView) findViewById(R.id.phoneString);
        refranceString = getIntent().getStringExtra(Tools.PLACE_REFRANCE);
        try {
            String localDetailString = MapsHttpUtil.getGetdPlaceDetailMessage(refranceString);
            JSONObject jsonObject = new JSONObject(localDetailString);
            JSONObject resultJsonObject = new JSONObject(jsonObject.getString("result"));
            placeName.setText("名称:" + resultJsonObject.getString("name"));
            placeAddress.setText("地址 :" + resultJsonObject.getString("formatted_address"));
            phoneNumber.setText("电话:" +
resultJsonObject.getString("formatted_phone_number"));
            mapURL = resultJsonObject.getString("url");
        } catch (Exception e) {
            e.printStackTrace();
        }}
    public void toGoogleMap(View view) {
        Intent i = new Intent(Intent.ACTION_VIEW, Uri.parse(mapURL));
        startActivity(i);
    }}
```

8.3.5 线路规划

因为在正式版本的 Android SDK 中，删除了原有 M5 版本里的 DrivingDirection package，所以无法通过程序来规划导航路线。虽然无法自行设计导航路线，但却可以调用手机内置的地图程序来传递导航坐标规划路径。

程序一开始调用 getLocationProvider()取得现有的 Location，以此取得目前所在位置的地理坐标（fromGeoPoint），而在页面设置中提供一个 EditText Widge 来让用户输入要前往的地址，通过地址反查目的地的地址坐标（toGeoPoint），再通过 Internet 的方式调用内置地图程序。

自定义方法 getGeoByLocation()传入的参数为 Location 对象，并返回该 Location 的 GeoPoint 对象，调用 LocationManager.requestLocationUpdates()设计 LocationListener，这个监听器在 GPS 打开作用下才会触发事件，在手机位置改变的同时，去的当下的 Location。路径规划 Intent 要打开的其实就是 URL 的类型，以 Uri。Parse()传入 Google Map 路线规划的网址：http://maps.google.com/maps?f=d,表示要使用 Google Map 的路径规划功能，以 saddr 为"起点"的经纬度，daddr 为"终点"经纬度。

线路规划代码如下。

```java
public class WayActivity extends MapActivity{
    protected void onCreate(Bundle savedInstanceState)
    {
        super.onCreate(savedInstanceState);
        /*创建LocationManager对象取得系统LOCATION服务*/
        mLocationManager01=(LocationManager)getSystemService(Context.LOCATION_SERVICE);
        //自定义方法，访问Location Provider
        getLocationProvider();
        /*传入Location对象，显示于MapView*/
        fromGeoPoint=getGeoByLocation(mLocation01);
        refreshMapViewByGeoPoint(fromGeoPoint, mMapView01,intZoomLevel);
        // 创建LocationManager对象，监听 Location更改时事件，更新MapView
        mLocationManager01.requestLocationUpdates(strLocationProvider, 2000, 10, mLocationListener01);
        mLocationManager01.requestLocationUpdates
        (strLocationProvider, 2000, 10, mLocationListener01);
        mButton01=(Button)findViewById(R.id.btnsetway);
        mButton01.setOnClickListener(new Button.OnClickListener()
        {public void onClick(View v)
            {
                if(mEditText01.getText().toString()!="")
                {/*取得User要前往地址的GeoPoint对象*/
                    toGeoPoint=
                        getGeoByAddress(mEditText01.getText().toString());
                    /*线路规划Intent*/
                    Intent intent=new Intent();
                    intent.setAction(android.content.Intent.ACTION_VIEW);
                    /*传入路径规划所需要的地标地址*/
                    intent.setData
                    (
                        Uri.parse("http://maps.google.com/maps?f=d&aaddr="+
                        GeoPointToString(fromGeoPoint)+
                        "&daddr="+GeoPointToString(toGeoPoint)+
                        "&h1=cn")
                    );
                    startActivity(intent);
```

```java
        }}});
    private final LocationListener mLocationListener01 = new LocationListener()
    {
        public void onLocationChanged(Location location)
        {
            /*当手机收到位置更改时,将Location传入getMyLocation*/
            mLocation01=location;
            fromGeoPoint=getGeoByLocation(location);
            refreshMapViewByGeoPoint(fromGeoPoint,mMapView01,intZoomLevel);
        }
    /*传入Location对象,取回其GeoPoint对象*/
    private GeoPoint getGeoByLocation(Location location) {
        GeoPoint gp=null;
        try
        { /*当Location存在*/
            if(location!=null)
            {   double geoLatitude=location.getLatitude()*1E6;
                double geoLongitude=location.getLongitude()*1E6;
                gp=new GeoPoint((int)geoLatitude,(int)geoLongitude);
            }}
        catch(Exception e)
        {e.printStackTrace();}
        return gp;}
    /*输入地址,取得其GeoPoint对象*/
    private GeoPoint getGeoByAddress(String strSearchAddress) {
        GeoPoint gp=null;
        try
        {   if(strSearchAddress!="")
            {   Geocoder mGeocoder01=new Geocoder
                    (WayActivity.this,Locale.getDefault());
                List<Address> lstAddress=mGeocoder01.getFromLocationName
                    (strSearchAddress,1);
                if(!lstAddress.isEmpty())
                {   Address adsLocation=lstAddress.get(0);
                    double geoLatitude=adsLocation.getLatitude()*1E6;
                    double geoLongitude=adsLocation.getLongitude()*1E6;
                    gp=new GeoPoint((int)geoLatitude,(int)geoLongitude);
```

```java
        }}}
    catch(Exception e)
    {   e.printStackTrace();}
    return gp;
}
/*传入geoPoint更新MapView里的Google Map*/
private void refreshMapViewByGeoPoint(GeoPoint gp,
        MapView mapview, int zoomLevel) {
    try
    {   mapview.displayZoomControls(true);
        MapController myMC=mapview.getController();
        myMC.animateTo(gp);
        myMC.setZoom(zoomLevel);
        mapview.setSatellite(false);}
    catch(Exception e)
    {   e.printStackTrace();}
}
/*传入经纬度更新MapView里的Google Map*/
public static void refreshMapViewByCode
(double latitude, double longitude,
        MapView mapview, int zoomLevel)
{   try
    {   GeoPoint p=new GeoPoint((int)latitude,(int)longitude);
        mapview.displayZoomControls(true);
        MapController myMC=mapview.getController();
        myMC.animateTo(p);
        myMC.setZoom(zoomLevel);
        mapview.setSatellite(false);
    }
    catch(Exception e)
    {   e.printStackTrace();}
}
/*将GeoPoint里的经纬度以String， String返回*/
private String GeoPointToString(GeoPoint gp) {
    String strReturn="";
    try
    {   /*当Location存在*/
```

Android 应用程序设计

```
                if(gp!=null)
            {   double geoLatitude=(int)gp.getLatitudeE6()/1E6;
                double geoLongitude=(int)gp.getLongitudeE6()/1E6;
                strReturn=String.valueOf(geoLatitude)+","+
                String.valueOf(geoLongitude);}}
        catch(Exception e)
        {    e.printStackTrace();}
        return strReturn;}
/*取得LocationProvider*/
private void getLocationProvider() {
    try
    {   Criteria mCriteria01=new Criteria();
        mCriteria01.setAccuracy(Criteria.ACCURACY_FINE);
        mCriteria01.setAltitudeRequired(false);
        mCriteria01.setBearingRequired(false);
        mCriteria01.setCostAllowed(true);
        mCriteria01.setPowerRequirement(Criteria.POWER_LOW);
        strLocationProvider=
                mLocationManager01.getBestProvider(mCriteria01,true);
        mLocation01=mLocationManager01.getLastKnownLocation(strLocationProvider);
    }
    catch(Exception e)
    {   mTextView01.setText(e.toString());
        e.printStackTrace();}}}
```

8.3.6 分享给好友

"分享给好友"功能通过 ShareActivity.java 实现。关键源代码如下。

```
public class ShareActivity extends Activity{
    public void onCreate(Bundle savedInstanceState) {
        super.onCreate(savedInstanceState);
        /*加载main。xml Layout*/
        setContentView(R.layout.sharelayout);
      /*实现分享*/
    Button btn_share=(Button)findViewById(R.id.btn_share);
    btn_share.setOnClickListener(new Button.OnClickListener()
    {    public void onClick(View v)
```

232

{   Intent intent=**new** Intent(Intent.*ACTION_SEND*);
    intent.setType("image/*");
    intent.putExtra(Intent.*EXTRA_SUBJECT*,"分享");
    intent.putExtra(Intent.*EXTRA_TEXT*,"终于可以了");
    intent.setFlags(Intent.*FLAG_ACTIVITY_NEW_TASK*);
    startActivity(Intent.*createChooser*(intent,getTitle()));}}); }}

8.4 项目成果（见图 8-6～图 8-9）

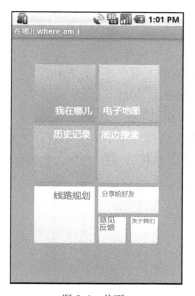

图 8-6　首页

图 8-7　电子地图

图 8-8　路径规划

图 8-9　地图上搜索

参考教材

[1] 杨丰盛.Android 应用开发揭秘[M]．北京：机械工业出版社，2010.

[2] 李宁.Android 开发权威指南[M]．北京：人民邮电出版社，2011.

[3] 盖索林. Google Android 开发入门指南（第 2 版） [M]．北京：人民邮电出版社，2009.

[4] 梅尔著，王超译. Android 2 高级编程（第 2 版）（英）[M]．北京：清华大学出版社，2010.

[5] 余志龙等著，王世江改编. Google Android SDK 开发范例大全(第 2 版)[M]．北京：人民邮电出版社，2010.

[6] 李宁.Android/OPhone 开发完全讲义[M]．北京：水利水电出版社，2010.

[7] 李刚. 疯狂 Android 讲义[M]．北京：电子工业出版社，2011.

[8] 汪永松. Android 平台开发之旅[M]．北京：机械工业出版社，2010.

[9] E2EColud 工作室. 深入浅出 Google Android[M]．北京：人民邮电出版社，2009.

[10] 梅尔著，王鹏杰，霍建同译. Android 高级编程.(英)[M]．北京：清华大学出版社，2010.